T0297107

DISASTER EPIDEMIOLOGY

DISASTER EPIDEMIOLOGY

METHODS AND APPLICATIONS

Edited by

Jennifer A. Horney

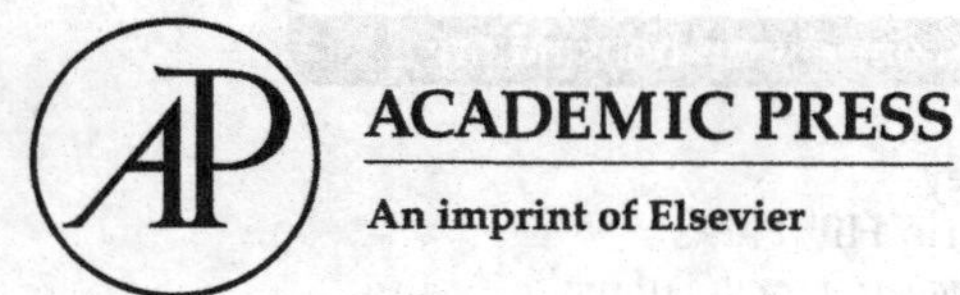

Academic Press is an imprint of Elsevier
125 London Wall, London EC2Y 5AS, United Kingdom
525 B Street, Suite 1800, San Diego, CA 92101-4495, United States
50 Hampshire Street, 5th Floor, Cambridge, MA 02139, United States
The Boulevard, Langford Lane, Kidlington, Oxford OX5 1GB, United Kingdom

Library of Congress Cataloging-in-Publication Data
A catalog record for this book is available from the Library of Congress

British Library Cataloguing-in-Publication Data
A catalogue record for this book is available from the British Library

ISBN: 978-0-12-809318-4

For information on all Academic Press publications visit our website at https://www.elsevier.com/books-and-journals

Publisher: Mica Haley
Acquisition Editor: Erin Hill-Parks
Editorial Project Manager: Tracy Tufaga
Production Project Manager: Kiruthika Govindaraju
Designer: Christian Bilbow

Typeset by TNQ Books and Journals

To my husband, Michael, and my children, Ian and Lila. In memory of my beloved father, Joseph Horney, who was resolute in his support for all my pursuits, both personal and professional.

Contents

5. Applications: Disaster-Related Mortality Surveillance
REBECCA S. NOE

6. Methods: Study Designs in Disaster Epidemiology
LATASHA A. ALLEN, JENNIFER A. HORNEY

7. Applications: Community Assessment for Public Health Emergency Response
AMY H. SCHNALL, AMY WOLKIN, TESFAYE M. BAYLEYEGN

8. Applications: Assessment of Chemical Exposures
MAUREEN F. ORR, MARY ANNE DUNCAN†

Vignette: Geothermal Venting and Emergency Preparedness—Lake County, California
SVETLANA SMORODINSKY, TRACY BARREAU, JASON WILKEN

Vignette: CASPER in Response to a Slow-Moving Disaster—The California Drought
SVETLANA SMORODINSKY, JASON WILKEN, TRACY BARREAU

9. Methods: Questionnaire Development and Interviewing Techniques
AMY H. SCHNALL, AMY WOLKIN, NICOLE NAKATA

Vignette: Investigating Foodborne Outbreaks
MARILYN FELKNER, VENESSA CANTU, KEVIN MCCLARAN

10. Applications: Social Vulnerability to Disaster (Hampton and Hertford Counties—Isabel)
DANIELLE SPURLOCK

11. Applications: Emergency Responder Health Monitoring and Surveillance

RENÉE H. FUNK

Vignette: Experiences Working With Ebola Virus Disease and Pregnancy in Sierra Leone, 2014

JONETTA JOHNSON MPOFU, MICHELLE DYNES

12. Methods: Data Analysis for Disaster Epidemiology

ASHLEY CONLEY

13. Applications: Biosurveillance, Biodefense, and Biotechnology

KOYA C. ALLEN

Vignette: Climate Change Effects on Flooding During Hurricane Sandy (2012)

JOEL C. DIETRICH

Vignette: Disasters and Chronic Medical Conditions

PAMELA ALLWEISS

14. Applications: Disaster Communication and Community Engagement

JENNIFER C. BEGGS

Vignette From Recent Responses: Roseburg, Oregon Mass Shooting

AKIKO M. SAITO

15. What Can Disaster Epidemiology Contribute to Building Resilient Communities?

JENNIFER A. HORNEY

List of Contributors

Koya C. Allen US Department of Defense, Stuttgart, Germany

Latasha A. Allen Office of the Assistant Secretary for Preparedness and Response (ASPR), Office of Emergency Management (OEM), Washington, DC, United States

Pamela Allweiss Centers for Disease Control and Prevention, Atlanta, GA, United States

Tracy Barreau Centers for Disease Control and Prevention, Atlanta, GA, United States, U.S. Public Health Service Commissioned Corps, Rockville, MD, United States; California Department of Public Health, Richmond, CA, United States

Tesfaye M. Bayleyegn Centers for Disease Control and Prevention, Atlanta, GA, United States

Jennifer C. Beggs Michigan Department of Health and Human Services, Lansing, MI, United States

Venessa Cantu Texas Department of State Health Services, Austin, TX, United States

Karen Chu VA Greater Los Angeles Healthcare System, Sepulveda, CA, United States

Ashley Conley St. Joseph Hospital, Nashua, NH, United States

Joel C. Dietrich NC State University, Raleigh, NC, United States

Hope Dishman Georgia Department of Public Health, Atlanta, GA, United States

Aram Dobalian University of California, Los Angeles, CA, United States; VA Greater Los Angeles Healthcare System, Sepulveda, CA, United States

Mary Anne Duncan† Agency for Toxic Substances and Disease Registry, Atlanta, GA, United States

Michelle Dynes Centers for Disease Control and Prevention, Atlanta, GA, United States; US Public Health Service Commissioned Corps, Rockville, MD, United States

Laura Edison Georgia Department of Public Health, Atlanta, GA, United States; Centers for Disease Control and Prevention, Atlanta, GA, United States

Marilyn Felkner Texas Department of State Health Services, Austin, TX, United States

Renée H. Funk Centers for Disease Control and Prevention, Atlanta, GA, United States

Rebecca J. Heick Augustana College, Rock Island, IL, United States

Jennifer A. Horney Texas A&M University, College Station, TX, United States

Josephine Malilay Centers for Disease Control and Prevention, Atlanta, GA, United States

Kevin McClaran Texas Department of State Health Services, Austin, TX, United States

Jonetta Johnson Mpofu Centers for Disease Control and Prevention, Atlanta, GA, United States; US Public Health Service Commissioned Corps, Rockville, MD, United States

Nicole Nakata Centers for Disease Control and Prevention, Atlanta, GA, United States

Rebecca S. Noe Centers for Disease Control and Prevention, Atlanta, GA, United States

Maureen F. Orr Agency for Toxic Substances and Disease Registry, Atlanta, GA, United States

†Deceased

Tiffany Radcliff Texas A&M University, College Station, TX, United States; VA Greater Los Angeles Healthcare System, Sepulveda, CA, United States

Akiko M. Saito Oregon Health Authority – Public Health Division, Portland, OR, United States

Amy H. Schnall Centers for Disease Control and Prevention, Atlanta, GA, United States

Suzanne Shurtz Texas A&M University, College Station, TX, United States

Kanta Sircar Centers for Disease Control and Prevention, Chamblee, GA, Unites States

Svetlana Smorodinsky California Department of Public Health, Richmond, CA, United States

Karl Soetebier Georgia Department of Public Health, Atlanta, GA, United States

Danielle Spurlock University of North Carolina, Chapel Hill, NC, United States

Dorothy Stearns Centers for Disease Control and Prevention, Chamblee, GA, Unites States

Kahler Stone Texas A&M University, College Station, TX, United States

Jason Wilken Centers for Disease Control and Prevention, Atlanta, GA, United States, U.S. Public Health Service Commissioned Corps, Rockville, MD, United States; California Department of Public Health, Richmond, CA, United States

Amy Wolkin Centers for Disease Control and Prevention, Atlanta, GA, United States

Acknowledgments

This book is the result of important contributions by a large number of public health practitioners and academicians, and the resulting combination of both epidemiologic methods and applications has been a critical focus of the book from its inception. For the last decade or more, a dedicated group of experts in epidemiology and environmental health have led the charge to recognize the importance of disaster epidemiology within the fields of public health and epidemiology, as well as within the phases of the disaster management cycle. This book would not have been possible without the work of many people who are not listed as authors, including members of the Center for Disease Control and Prevention's Disaster Epidemiology Community of Practice (DECoP), formerly known as the Disaster Surveillance Workgroup, the Council of State and Territorial Epidemiologists' Disaster Epidemiology Subcommittee, and attendees at the annual National Disaster Epidemiology Workshop. I would particularly like to acknowledge the many applied public health practitioners at the federal, state, and local levels who have shared their time and their real-world expertise with me and given me an opportunity to learn what cannot be taught in a classroom. Most importantly, these practitioners have been generous with my students, building on what I can teach them, and helping them become the next generation of disaster epidemiologists.

Introduction to Disaster Epidemiology

Josephine Malilay[1], *Jennifer A. Horney*[2]

[1]**Centers for Disease Control and Prevention, Atlanta, GA, United States;**
[2]**Texas A&M University, College Station, TX, United States**

INTRODUCTION

Despite efforts to control, prevent, and mitigate public health consequences of natural disasters and complex emergencies in recent decades, these events continue to affect increased numbers of people worldwide and lead to a significant number of deaths, injuries, diseases, and disabilities (Leaning & Guha Sapir, 2013). The study of the health impacts of disasters and emergencies has led to an expanding use of methods borrowed from epidemiology, the science that studies the causes and determinants of diseases in populations (Last, 1988). The use of epidemiologic methods and study designs in postdisaster settings can aid in identifying vulnerable populations, quantifying disaster-related morbidity and mortality, determining effects of a disaster event on predisaster health status, and informing decision-making for appropriate interventions and allocation of resources for relief, recovery, and resilience building programs.

The application of epidemiologic methods in the disaster setting has led to the emergence of "disaster epidemiology," which began with the study of health risks after environmental disasters and is now widely regarded as a subset of environmental epidemiology (Lauriola & Leonardi, 2016). Beginning with descriptive techniques, one of the earliest disaster epidemiology applications was an evaluation of health status after a cyclone struck East Bengal, Bangladesh in 1970. Two postdisaster assessments identified young children, elderly, and females as vulnerable subgroups of a densely populated and impoverished region and provided reasonably accurate estimates of disaster relief requirements so that the region could recover agricultural self-sufficiency (Sommer & Mosley, 1972). Since then, other public health pioneers with years of field experience have pointed out the utility of epidemiologic methods in addressing the health effects of disasters (Lechat, 1976; Seaman, 1984) and described the role of the epidemiologist in natural disasters (Binder & Sanderson, 1987). In 2014, after numerous applications of the use of epidemiologic methods including rapid needs assessments in diverse field settings, the United States–based Council for State and Territorial Epidemiologists (CSTE) and the US Centers for Disease Control and Prevention (CDC) published a conceptual framework describing the role of disaster epidemiology and outlining specific epidemiologic activities that could be conducted during different phases of the disaster management cycle (Malilay et al., 2014). The framework provides guidance for using epidemiologic methods in investigations and studies during preparedness, response, recovery, and mitigation. However, the process during the preceding four and half decades deserves mention.

DESCRIPTIVE STUDIES

Initially, disaster epidemiology studies were conducted after certain events, often as a result of the intervention of outside agencies or groups due to the magnitude of disaster. Occasionally, natural experiments took place in areas where ongoing predisaster studies allowed for timely response by seamless allocation of resources and staff to postdisaster studies. Many descriptive studies described mortality and morbidity related to the event. For instance, counts of fatalities are summarized in the literature for specific events including earthquakes, volcanic eruptions, hurricanes, tornadoes, and floods (Baxter et al., 1981; Brenner & Noji, 1992; CDC, 1985, 1989, 1990, 1993a,b,c, 1994; Glass et al., 1980; Peek-Asa et al., 1998; Rosenfield, McQueen, & Lucas, 1994). Fatalities and injuries were largely attributed to trauma and traumatic rhabdomyolysis from crush injuries in earthquakes (CDC, 1990; Peek-Asa et al., 1998), described by their location on the body (e.g., head, clavicle, extremities) in tornadic events (Brenner & Noji, 1992; Brown, Archer, Kruger, & Mallonee, 2002; Glass et al., 1980; Rosenfield et al., 1994), or explained by individual activity during preparation or cleanup (e.g., electrocution while securing outside electrical items or in postflood cleanup work) (CDC, 1993b, 1994). In some volcanic disasters, a variety of health outcomes were observed, ranging from respiratory conditions from ashfall-related hazards to burns from tephra associated with explosive eruptions (Baxter, Bernstein, Falk, French, & Ing, 1982). In hydrometeorological events such as hurricanes and flash flooding, fatalities tend to be mostly related to drowning, and in particular, drowning among occupants of motor vehicles where drivers attempted to drive through flood waters of unknown depth (CDC, 1993a, 1993b, 1994). In tornadoes, fatalities were mainly attributed to trauma (Brenner & Noji, 1992; Brown et al., 2002; Edison, Lybarger, Parsons, Maccormack, & Freeman, 1990; Glass et al., 1980; Rosenfield et al., 1994), while the health impacts from major winter storms included carbon monoxide poisoning due to the inappropriate use of generators (Daley et al., 2005).

Issues Related to Methods

The use of descriptive techniques prompted questions of whether a fatality, injury, or other health outcome could be causally attributed to a disaster event, and more specifically, directly or indirectly related to the mechanical forces of the disaster event, considered to be the exposure. Efforts were made to involve stakeholders such as the National Oceanic and Atmospheric Administration (NOAA) and the National Association of Medical Examiners in establishing causal distinctions and a working matrix was proposed (Combs, Quenemoen, Parrish, & Davis, 1999). Today, as mentioned previously, collaborative work to facilitate the distinction is led by the CSTE and CDC. Other methodological advances include the development of a new reporting form to identify disaster-related deaths proposed by the National Center for Health Statistics (National Center for Health Statistics, 2016). By using these and other tools, commonalities and differences among similar disaster types are now observed, setting the stage for the application of analytic techniques to ascertain risks for individual behaviors and other pertinent factors in various post-disaster events.

RISK INVESTIGATIONS AND STUDIES

Characterization of the risks in some types of disaster events has developed. For example, in a classic study of earthquake impacts in Guatemala in 1976, Glass et al. (1977) examined structural characteristics, as well as individual characteristics and behaviors among occupants of earthquake impacted structures. Subsequent studies in postearthquake settings examined these issues and sought to identify floors within multistoried structures with the greatest chance of survival (Armenian, Noji, & Oganesian, 1992; Peek-Asa et al., 1998; Roces, White, Dayrit, & Durkin, 1992). Results suggested areas where search and rescue teams could prioritize a search as the top, middle, or bottom of rubble. In other postdisaster settings, risks for carbon monoxide poisoning after a major ice storm pointed to improper use of fuel-powered generators when electrical power was out (Daley, Smith, Paz-Argandona, Malilay, & McGeehin, 2000; Lutterloh et al., 2011). Other analytic studies in tornado, tropical cyclone, and landslide settings addressed the utility of forecasting and warning systems, access and use of NOAA weather radios and receipt of messages, and occupant behavior on hearing of a rapidly moving weather event, such as a tornado (Brown et al., 2002; Daley et al., 2005; Hammer & Schmidlin, 2002; Hayden et al., 2007; Paul, Stimers, & Caldas, 2015; Perreault, Houston, & Wilkins, 2014). Results from each of these studies add to a growing body of knowledge related to the appropriate actions to take for shelter seeking in a tornado event.

Although the current literature continues to grow, much is yet to be done in terms of gathering baseline data for comparison. For example, case-control studies document appropriate behaviors for shelter seeking in high-rise structures during seismic events but are not typically generalizable for earthquake-prone regions outside of the study area (Armenian et al., 1992; Roces et al., 1992). Results from tornado studies indicate that risks for injury and death are greater when attempting to outrun a tornado in a vehicle or remaining in a vehicle (Glass et al., 1980); however, attempting to outrun a tornado also has been shown to be protective (CDC, 1988; Daley et al., 2005). Additional studies of risk and protective factors with appropriate considerations for related infrastructure characteristics, evacuation behaviors, and potential confounders are needed to build the evidence for formulating guidelines and recommendations.

RAPID NEEDS ASSESSMENTS

An almost immediate response activity is the requirement to determine health-related needs in disaster-affected populations. Although a number of approaches have been developed to address this challenge, one method that has gained wide acceptance in the state and local governments in the United States is the Community Assessment for Public Health Emergency Response, known as CASPER. First developed in the 1960s by the World Health Organization to estimate immunization coverage in an unknown population, this rapid and cost-effective technique has evolved to become a useful tool to estimate population needs in affected communities (CDC, 2012). The method now employs population data from the U.S. Census and geographic information systems such as the ArcGIS platform

(ESRI, Redlands, California) to select clusters and households for the survey. The rapid availability of results from CASPER has encouraged further application in nondisaster settings, such as use of health services and vaccine coverage in a given population (Bayleyegn et al., 2015; Horney, Davis, Davis, & Fleischauer, 2013; Horney, Moore, Davis, & MacDonald, 2010). Also, repeated application of the method can allow emergency officials to observe any trends or changes over time (Kirsch et al., 2016).

Surveillance

Surveillance systems and techniques have also been developed or adapted for use as disaster epidemiology tools in various settings. For example, to quell rumors of epidemics of infectious disease and transmission by cadavers (PAHO, 1994; Morgan, 2004), surveillance systems were instituted to provide information about the health status of displaced populations, populations remaining in the given area, and workers assisting in the relief efforts. A characterization of death and injuries or selected diseases and disabilities over a time period of interest can be presented using epidemic curves. Baseline or comparative information may be obtained by extracting similar data for the same time period from previous years; differences and similarities for the disaster study period are compared with baseline values for a given area. In some events, proportionate morbidity techniques are applied to compare prevalence of one disease with another. Active and passive systems using various sources of data such as hospitals and health-care facilities can be established postdisaster. Surveillance in shelters for evacuees or camps for displaced people can be established. For example, postdisaster surveillance identified an increase in the number of cases of coccidioidomycosis related to fungus-contaminated dust storms from the Northridge earthquake in California (Schneider et al., 1997), gastrointestinal illness associated with the Midwestern floods (Wade et al., 2004), norovirus in shelters after Hurricane Katrina (CDC, 2005), and cholera several months after a devastating earthquake in Haiti (Barzilay et al., 2013). Clearly, the need to monitor health conditions through surveillance systems remains a requisite epidemiologic response activity.

REGISTRIES AND ACCOUNTABILITY STUDIES

While still relatively new, the use of registries, particularly for long-term studies to address potential chronic health effects of disasters, allows for a more careful long-term follow-up of people affected by a disaster event. A registry instituted after the World Trade Center disaster in 2001 led to the identification of cancers from exposures to ignited chemicals and material among firefighters and emergency responders (Li et al., 2012). The National Institute for Environmental Health Sciences (NIEHS) initiated a similar registry in the aftermath of the 2010 Deepwater Horizon oil spill to ascertain the health outcomes of exposed individuals involved in cleanup and remediation activities (Kwok et al., 2017). Accountability studies, also called evaluation studies, assess the efficacy of interventions such as relief programs. In one study, an evaluation of the effectiveness of mental health outreach teams after Hurricane Andrew in Florida found that the teams had no significant impact on mental distress or the use of mental health services (McDonnell et al., 1995).

SUMMARY AND CONCLUSIONS

At this juncture, we may ask "What are the next steps for disaster epidemiology?" To summarize, we have seen the evolution of community-based rapid needs assessments, the development and implementation of surveillance systems capturing infectious diseases and chronic health conditions, the conduct of descriptive and analytic studies addressing risk and protective factors for preventing and controlling adverse health outcomes, as well as applications of registries and accountability studies. New hypotheses may be formulated from the results of rapid needs assessments and surveillance and tested in more complex study designs. Information about expected behaviors in warning and forecasting systems and safe sheltering still need corroboration with additional studies. We expect the knowledge base provided by disaster epidemiology to grow exponentially as the body of evidence has only just begun to be compiled.

Disaster epidemiology can also expand along with advances in epidemiologic methods and in the discipline of environmental epidemiology. Longer-term exposures to potentially complex mixtures resulting from the combined effects of natural disasters and environmental contamination may be associated with new health outcomes, including those related to mental health and nutritional status. Electronic health records and the collection of information in real time may also affect current epidemiologic methods and subsequent response and recovery efforts. Large climatic datasets and data collected by citizen scientists are now generated by smartphones, sensors, and satellites and require hardware and software for processing as well as the application of novel analytic techniques. Finally, increased use of ongoing biomonitoring and environmental sampling, where appropriate, may assist in identifying and quantifying disaster-related exposures to chemical contaminants of concern.

Wide-scale disasters, particularly those whose effects may potentially be enhanced by climate change, may also reshape study designs and the interpretation of findings that address partial, cumulative, additive, or synergistic health effects as well as attributable risks and environmental impacts. The emerging field of integrative epidemiology may also require a composite of skills from various areas of epidemiology, including infectious diseases, reproductive health, and birth defects and developmental disabilities. For example, floods in an area where the Zika virus is endemic may require closer surveillance of shelter occupants of childbearing age, greater source reduction and vector mitigation efforts, and increased personal protective measures. Clearly, more can be done to build the evidence base from disaster epidemiology to improve disaster preparedness, response, recovery, and mitigation for all.

References

Armenian, H. K., Noji, E. K., & Oganesian, A. P. (1992). A case-control study of injuries arising from the earthquake in Armenia, 1988. *Bulletin of the World Health Organization, 70*(2), 251–257.

Barzilay, E. J., Schaad, N., Magloire, R., Mung, K. S., Boncy, J., Dahourou, G. A., et al. (2013). Cholera surveillance during the Haiti epidemic – the first 2 years. *The New England Journal of Medicine, 368*(7), 599–609.

Baxter, P. J., Bernstein, R. S., Falk, H., French, J., & Ing, R. (1982). Medical aspects of volcanic disasters: An outline of the hazards and emergency response measures. *Disasters, 6*(4), 268–276.

Baxter, P. J., Ing, R., Falk, H., French, J., Stein, G. F., Bernstein, R. S., et al. (1981). Mount St Helens eruptions, May 18 to June 12, 1980. An overview of the acute health impact. *JAMA, 246*(22), 2585–2589.

Bayleyegn, T. F., Schnall, A. H., Ballou, S. G., Zane, D. F., Burrer, S. L., Noe, R. S., et al. (2015). Use of community assessments for public health emergency response (CASPERs) to rapidly assess public health issues – United States, 2003–2012. *Prehospital and Disaster Medicine, 30*(4), 374–381.

Binder, S., & Sanderson, L. M. (1987). The role of the epidemiologist in natural disasters. *Annals of Emergency Medicine, 16*(9), 1081–1084.

Brenner, S. A., & Noji, E. K. (1992). Head and neck injuries from 1990 Illinois tornado. *American Journal of Public Health, 82*(9), 1296–1297.

Brown, S., Archer, P., Kruger, E., & Mallonee, S. (2002). Tornado-related deaths and injuries in Oklahoma due to the 3 May 1999 tornadoes. *Weather Forecast, 17*, 343–353.

Centers for Disease Control and Prevention. (1985). Epidemiologic notes and reports tornado disaster – North Carolina, South Carolina, March 28, 1984. *Morbidity and Mortality Weekly Report, 34*(15), 205–206, 211–213.

Centers for Disease Control and Prevention. (1988). Tornado disaster – Texas. *Morbidity and Mortality Weekly Report, 37*(30), 454–456, 461.

Centers for Disease Control and Prevention. (1989). Deaths associated with hurricane Hugo–Puerto Rico. *Morbidity and Mortality Weekly Report, 38*, 680–682.

Centers for Disease Control and Prevention. (1990). International notes earthquake disaster – Luzon, Philippines. *Morbidity and Mortality Weekly Report, 39*(34), 573–577.

Centers for Disease Control and Prevention. (1993a). Injuries and illnesses related to hurricane Andrew – Louisiana, 1992. *Morbidity and Mortality Weekly Report, 42*(13), 242–243, 250–251.

Centers for Disease Control and Prevention. (1993b). Flood-related mortality – Missouri, 1993. *Morbidity and Mortality Weekly Report, 42*(48), 941–943.

Centers for Disease Control and Prevention. (1993c). Public health consequences of a flood disaster – Iowa, 1993. *Morbidity and Mortality Weekly Report, 42*(34), 653–656.

Centers for Disease Control and Prevention. (1994). Flood-related mortality – Georgia, July 4–14, 1994. *Morbidity and Mortality Weekly Report, 43*(29), 526–530.

Centers for Disease Control and Prevention. (2005). Norovirus outbreak among evacuees from hurricane Katrina – Houston, Texas, September 2005. *Morbidity and Mortality Weekly Report, 54*(40), 1016–1018.

Centers for Disease Control and Prevention. (2012). *Community assessment for public health emergency response (CASPER) toolkit* (2nd ed.). Atlanta, GA: Centers for Disease Control and Prevention.

Combs, D. L., Quenemoen, L. E., Parrish, R. G., & Davis, J. H. (1999). Assessing disaster-attributed mortality: Development and application of a definition and classification matrix. *International Journal of Epidemiology, 28*(6), 1124–1129.

Daley, W. R., Brown, S., Archer, P., Kruger, E., Jordan, F., Batts, D., et al. (2005). Risk of tornado-related death and injury in Oklahoma, May 3, 1999. *American Journal of Epidemiology, 161*(12), 1144–1150.

Daley, W. R., Smith, A., Paz-Argandona, E., Malilay, J., & McGeehin, M. (2000). An outbreak of carbon monoxide poisoning after a major ice storm in Maine. *The Journal of Emergency Medicine, 18*(1), 87–93.

Edison, M., Lybarger, J. A., Parsons, J. E., Maccormack, J. N., & Freeman, J. I. (1990). Risk factors for tornado injuries. *International Journal of Epidemiology, 19*(4), 1051–1056.

Glass, R. I., Craven, R. B., Bregman, D. J., Stoll, B. J., Horowitz, N., Kerndt, P., et al. (1980). Injuries from the Wichita Falls tornado: Implications for prevention. *Science, 207*(4432), 734–738.

Glass, R. I., Urrutia, J. J., Sibony, S., Smith, H., Garcia, B., & Rizzo, L. (1977). Earthquake injuries related to housing in a Guatemalan village. *Science, 197*(4304), 638–643.

Hammer, B., & Schmidlin, T. W. (2002). Response to warnings during the 3 may 1999 Oklahoma city tornado: Reasons and relative injury rates. *Weather Forecast, 17*, 577–581.

Hayden, M. H., Drobot, S., Radil, S., Benight, C., Gruntfest, E. C., & Barnes, L. R. (2007). Information sources for flash flood warnings in Denver, CO and Austin, TX. *Environmental Hazards, 7*(3), 211–219.

Horney, J., Davis, M. K., Davis, S. H. E., & Fleischauer, A. (2013). An evaluation of community assessment for public health emergency response (CASPER) in North Carolina, 2003-2010. *Prehospital and Disaster Medicine, 28*(2), 94–98.

Horney, J. A., Moore, Z., Davis, M., & MacDonald, P. D. (2010). Intent to receive pandemic influenza A (H1N1) vaccine, compliance with social distancing and sources of information in NC, 2009. *PLoS One, 5*(6), e11226.

Kirsch, K. R., Feldt, B. A., Zane, D. F., Haywood, T., Jones, R. W., & Horney, J. A. (2016). Longitudinal community assessment for public health emergency response to wildfire, Bastrop County, Texas. *Health Security, 14*(2), 93–104.

Kwok, R. K., Engel, L. S., Miller, A. K., Blair, A., Curry, M. D., & Jackson, W. B. (2017). The GuLF STUDY: A prospective study of persons involved in the Deepwater Horizon oil spill response and clean-up. *Environmental Health Perspectives, 125*(4), 570.

Last, J. M. (1988). What is "clinical epidemiology?". *Journal of Public Health Policy, 9*(2), 159–163.

Lauriola, P., & Leonardi, G. (August 2016). *Epidemiological preparedness and response to environmental disasters.* Rome, Italy: International Society for Environmental Epidemiology. Available at http://www.isee2016roma.org/epidemiological-preparedness-and-response-to-environmental-disasters/.

Leaning, J., & Guha Sapir, D. (2013). Natural disasters, armed conflict, and public health. *The New England Journal of Medicine, 369*, 1836–1842.

Lechat, M. F. (1976). The epidemiology of disasters. *Proceedings of the Royal Society of Medicine, 69*(6), 421–426.

Li, J., Cone, J. E., Kahn, A. R., Brackbill, R. M., Farfel, M. R., Greene, C. M., et al. (2012). Association between World Trade Center exposure and excess cancer risk. *JAMA, 308*(23), 2479–2488.

Lutterloh, E. C., Iqbal, S., Clower, J. H., Spiller, H. A., Riggs, M. A., Sugg, T. J., et al. (2011). Carbon monoxide poisoning after an ice storm in Kentucky, 2009. *Public Health Reports, 126*(Suppl. 1), S108–S115.

Malilay, J., Heumann, M., Perrotta, D., Wolkin, A. F., Schnall, A. H., Podgornik, M. N., et al. (2014). The role of applied epidemiology methods in the disaster management cycle. *American Journal of Public Health, 104*(11), 2092–2102.

McDonnell, S., Troiano, R. P., Barker, N., Noji, E., Hlady, W. G., & Hopkins, R. (1995). Long-term effects of hurricane Andrew: Revisiting mental health indicators. *Disasters, 19*(3), 235–246.

Morgan, O. (2004). Infectious disease risks from dead bodies following natural disasters. *Revista Panamericana de Salud Publica, 15*(5), 307–312.

National Center for Health Statistics. (2016). *A reference guide for certification of deaths in the event of a severe weather, human-induced or chemical/radiological disaster*. Hyattsville: Maryland.

Organización Panamericana de la Salud. (1994). In *Las condiciones de salud en la Américas*. Pan American Health Org.

Paul, B. K., Stimers, M., & Caldas, M. (2015). Predictors of compliance with tornado warnings issued in Joplin, Missouri, in 2011. *Disasters, 39*(1), 108–124.

Peek-Asa, C., Kraus, J., Bourque, L. B., Vimalachandra, D., Yu, J., & Abrams, J. (1998). Fatal and hospitalized injuries resulting from the 1994 Northridge earthquake. *International Journal of Epidemiology, 27*(3), 459–465.

Perreault, M. F., Houston, J. B., & Wilkins, L. (2014). Does scary matter?: Testing the effectiveness of new National Weather Service tornado warning messages. *Communication Studies, 65*(5), 484–499.

Roces, M. C., White, M. E., Dayrit, M. M., & Durkin, M. E. (1992). Risk factors for injuries due to the 1990 earthquake in Luzon, Philippines. *Bulletin of the World Health Organization, 70*(4), 509–514.

Rosenfield, A. L., McQueen, D. A., & Lucas, G. L. (1994). Orthopedic injuries from the Andover, Kansas, tornado. *The Journal of Trauma, 36*(5), 676–679.

Schneider, E., Hajjeh, R. A., Spiegel, R. A., Jibson, R. W., Harp, E. L., Marshall, G. A., et al. (1997). A coccidioidomycosis outbreak following the Northridge, Calif, earthquake. *JAMA, 277*(11), 904–908.

Seaman, J. (1984). *Epidemiology of natural disasters*. Basel, Switzerland: Karger.

Sommer, A., & Mosley, W. H. (1972). East Bengal cyclone of November, 1970: Epidemiological approach to disaster assessment. *Lancet, 299*(7759), 1029–1036.

Wade, T. J., Sandhu, S. K., Levy, D., Lee, S., LeChevallier, M. W., Katz, L., et al. (2004). Did a severe flood in the Midwest cause an increase in the incidence of gastrointestinal symptoms? *American Journal of Empidemiology, 159*(4), 398–405.

CHAPTER

1

History of Disaster Epidemiology: 1960–2015

Jennifer A. Horney

Texas A&M University, College Station, TX, United States

BACKGROUND

Studies of the impacts of natural and man-made disasters have been undertaken since the 1950s and 1960s, growing out of the study of evacuation planning and research on wartime relocation following World War II, and later, concerns about nuclear attacks or accidents (Tierney, 2007). Early studies were primarily descriptive and based mostly on survey data collected following disasters. Surveys tended to rely on self-reported impacts and the time frames in which data were collected varied from immediately postdisaster to months after the event (Adams & Adams, 1984). These early studies were wide-ranging with regard to disaster type, including both acute (e.g., hurricanes, flash floods, and earthquakes) and more chronic types of emergencies (winter storms, droughts, and volcanic activity) (Adams & Adams, 1984; Erikson, 1976; Glass, O'Hare, & Conrad, 1979; Killian, 1954; Spencer, Romero, & Feldman, 1977).

From the beginning of disaster research, the methods used were "hardly distinguishable" from the methods used in everyday public health and social science research (Mileti, 1987, p. 69). However, the circumstances, under which the methods were employed, were very different and varied depending on the phase of the disaster process that was being studied. For example, immediately after a disaster, it may be more difficult to develop hypotheses, design reliable studies, recruit appropriate controls, or locate subjects (Killian, 1956). Delays associated with obtaining Institutional Review Board approval of research protocols may hinder the immediate collection of data in the postdisaster period (National Institutes of Health, n.d.). Research, particularly by those from outside the affected area who did not experience the disaster firsthand, can place perceived or actual burdens on individual respondents, communities, or systems during disaster response and recovery (Korteweg, van Bokhoven, Yzermans, & Grievink, 2010; Laska & Peterson, 2011).

Disaster Epidemiology
http://dx.doi.org/10.1016/B978-0-12-809318-4.00001-0

For these and other reasons, disaster epidemiology has historically been somewhat marginalized from the broader discipline of epidemiology. However, the role of epidemiologists in responding to disasters has expanded dramatically in the last two decades. Rapid Needs Assessment (RNA) methods, adapted from the World Health Organization's Expanded Program on Immunization (EPI) method of estimating immunization coverage, began to be implemented regularly after humanitarian emergencies, beginning with the famines in Somalia (Boss, Toole, & Yip, 1994), and after disasters, beginning with Hurricane Andrew (CDC, 1992; Hlady et al., 1994; Malilay, Flanders, & Brogan, 1996). Emergency department surveillance systems were implemented to capture illness, injury, and mortality data to quantify the health impacts of disasters such as floods in Louisiana (Ogden, Gibbs-Scharf, Kohn, & Malilay, 2001) and Hurricane Floyd in North Carolina (CDC, 2000). Other types of epidemiologic investigations, such as retrospective and prospective cohort studies, began to be implemented to study the long-term effects of disasters such as the terrorist attacks of September 11, 2001 (Savitz et al., 2008), the Indian Ocean Tsunami (Nishikiori et al., 2006), and the Deepwater Horizon oil spill (Stewart et al., 2011). The systematic use of the term "disaster epidemiology" since the 1990s has helped to establish the discipline as a formal subset of epidemiology and encourage its ongoing development (Malilay et al., 2014).

Disasters continue to be a major cause of morbidity and mortality. In spite of large efforts to reduce the impacts and costs of natural disasters, in the United States, average annual federal expenditures to fund rebuilding from catastrophic and chronic losses have been rising faster than either population or gross domestic product (GDP) (Gall, Borden, Emrich, & Cutter, 2011). Globally, the number of people at risk continues to grow along with the populations of megacities located in vulnerable coastal areas, as does the cost of disasters relative to real global GDP (Bouwer, Crompton, Faust, Höppe, & Pielke, 2007). These trends will likely accelerate along with the destructive potential of natural disasters due to climate change and sea-level rise. The following sections present a review of the development in the field of disaster epidemiology and provide examples of disaster epidemiology investigations.

1960s–1980s

Rapid Needs Assessments

Postdisaster RNAs are an important way to mitigate adverse health effects of a disaster among a population. The RNA cluster sampling methods, initially developed by the World Health Organization's EPI in the 1970s, were initially used to obtain data about vaccine coverage and ensure availability of vaccines to all children globally by 1990 (Lemeshow & Robinson, 1985). Since data on immunization coverage and the burden of vaccine-preventable diseases (e.g., tetanus, diphtheria, measles, whooping cough, and polio) were not available, cluster sampling methods were utilized to collect rapid data on vaccination coverage. The data gathered from these initial assessments helped to identify gaps in vaccination coverage and provide data to support the implementation of new programs to address them (Table 1.1).

In the 1980s, the RNA cluster sampling method was used to collect information about breastfeeding and child nutrition to supplement available vital statistics and hospital records

TABLE 1.1 Vaccine Coverage Rates by WHO Region, Development Status, and Income Status, 1980

	Vaccine						
	BCG	DTP1	DTP3	MCV1	PAB	POL3	RCV1
WHO REGIONS							
African	9	8	5	6	2	6	0
American	53	68	50	51	3	67	20
Eastern Mediterranean	18	36	18	15	1	20	0
European	18	73	66	59	0	75	4
Southeast Asian	12	22	7	0	16	3	0
Western Pacific	6	12	8	5	6	5	0
DEVELOPMENT STATUS							
Developed economy	35	67	60	51	0	59	30
Developing	16	25	12	8	10	15	0
Economy in transition	4	93	90	90	0	90	0
Least developed	10	11	5	6	3	5	0
INCOME STATUS							
High income	38	66	57	49	18	55	26
Low income	16	15	9	10	3	10	0
Middle income	14	23	11	7	9	13	0
NA	3	93	90	90	0	90	0

BCG, bacille Calmette-Guérin (vaccine against tuberculosis); *DTP1*, first dose of diphtheria and tetanus toxoid with pertussis vaccine; *DTP3*, third dose of diphtheria and tetanus toxoid with pertussis vaccine; *HepB3*, third dose of hepatitis B vaccine; *Hib3*, third dose of Haemophilus influenza type B vaccine; *MCV*, measles-containing vaccine; *PAB*, protection at birth against tetanus; *Pol3*, third dose of polio vaccine.
WHO/UNICEF estimates of immunization coverage 1980.

(Frerichs & Tar, 1988a). Researchers also used cluster sampling methods to conduct a survey of 396 children in rural Myanmar, collecting data on the proportion of children immunized against diphtheria, pertussis, and tetanus (DPT) polio (OPV), and tuberculosis (BCG) and the ages of children when mothers stopped breastfeeding and when children were first fed protein-rich foods (Table 1.2) (Frerichs & Tar, 1988b). Data were collected on a portable computer and analyzed shortly after interviews were completed; meaning that the time between data collection and the delivery of a report to the Ministry of Health was only eight days (Frerichs & Tar, 1988b).

Public Health Surveillance and Tracking Systems

During the 1960s, most countries had not developed a sufficient public health infrastructure to support regular public health surveillance activities. Systems that did exist focused on

TABLE 1.2 Dietary Intake Among Children in Rural Burma by Age, 1987

	Age in Years		
	0 to 1	1 to 2	2 to 3
Breast milk only	115	18	2
Breast milk and solids[a]	31	89	44
Solids[a] only	2	17	72
Neither	2	0	4

[a]*Solids include eggs, fish, or meat.*
Adapted from Frerichs, R. R., & Tar, K. T. (1988a). Breastfeeding, dietary intake and weight-for-age of children in rural Burma. Asia Pacific Journal of Public Health, 2, 16–21.

communicable diseases such as malaria, influenza, yellow fever, dengue, tuberculosis, and cholera; however, little microbiological laboratory capacity was available to support these surveillance programs (Raska, 1966). Even in low resource settings, the need for postdisaster epidemiologic surveillance was made clear after a major earthquake killed 23,000 and injured 77,000 in Guatemala in February, 1976. The Ministry of Health's surveillance network was severely affected by the destruction of health facilities and the loss of communication and transportation infrastructure (Spencer et al., 1977). The lack of public health surveillance data was having a major impact on decisions related to opening field hospitals, conducting vaccination campaigns, and controlling rumors of epidemics (Spencer et al., 1977). Even with limited capacity, a surveillance system was established to collect data on cases of selected diseases, the number of available hospital beds, the availability of medications, and potential outbreaks.

Following a major winter storm that impacted the Northeastern United States in February, 1978, surveillance of emergency rooms and town clerks was established to track morbidity and mortality related to the storm (Glass et al., 1979). The surveillance system allowed public health officials to provide evidence that disease outbreaks and environmental contamination events were not occurring, and therefore quell widely circulating rumors.

Epidemiology Investigations and Studies

During this time period, epidemiological investigations and studies relied heavily on surveys implemented using convenience, simple random, or cluster sampling. For example, after a major cyclone struck Bangladesh in November, 1970, Sommer and Mosley (1972) used cluster sampling to conduct an assessment of cyclone-related morbidity, nutritional status, housing quality, and access to drinking water and latrines over a period of several months. Some epidemiologic studies were also implemented to address the potential long-term impacts of disasters on health. For example, five years after a major flooding disaster associated with Hurricane Agnes (Fig. 1.1), factors such as property loss, financial difficulties, use of alcohol, and perceived distress during the recovery period were associated with continued high blood pressure, somatization, and anxiety among residents of an affected town in Pennsylvania (Logue, Hansen, and Hansen 1980).

FIGURE 1.1 Hurricane Agnes associated flooding in Farmville, Virginia. *Photo Credit: No. 72-893, Virginia Governor's Negative Collection, Courtesy of the Library of Virginia.*

Beyond natural disasters, in a review of industrials disasters occurring in the 1980s, Bertazzi (1989) outlined a number of epidemiological investigations that were conducted through the use of surveys to ascertain the immediate impacts on the health of affected populations. In addition to postdisaster questionnaires, Bertazzi suggested that follow-up studies should be implemented to better understand longer-term health impacts and that these studies should include the collection and maintenance of biological samples to evaluate the initial extent of exposure as well as changes over time. For example, multiple follow-up studies were conducted by the Bhopal Gas Disaster Research Center and the Indian Council for Medical Research following exposure to methyl isocyanate gas that leaked from a Union Carbide plant in Bhopal, India, in 1984, to enable tracking of long-term mortality and morbidity (Dhara & Dhara, 2002). Multiple cohorts, including cleanup workers and children born to mothers exposed to radiation from the 1986 meltdown of the Chernobyl Nuclear Power Plant in Ukraine, continue to be followed to monitor for health outcomes such as suicide and cancers (Ivanov et al., 2012; Rahu, Rahu, Tekkel, & Bromet, 2006).

1990s

Rapid Needs Assessments

Epidemiological approaches to disaster management and response evolved and grew significantly during the 1990s, with the implementation of better emergency networks for disaster relief and more robust disaster planning (Lillibridge, Noji, & Burkle, 1993). At the same time, increased media attention to disaster relief efforts, as well as the growth of global communication networks put more pressure on disaster response agencies to identify potential public health issues sooner and rapidly implement interventions to mitigate the health consequences of disasters in affected communities. Even with advancing computing technologies, rapid assessments were still frequently conducted using nonprobability sampling methods. The level of resources required to conduct RNAs remained relatively high, typically requiring outside assistance from various experts to develop assessment instruments, conduct interviews, and analyze data. However, during the 1990s, advances—led by WHO and CDC—were made to demonstrate the use of modified cluster sampling to perform RNAs (Hlady et al., 1994; Malilay et al., 1996; Turner, Magnani, & Shuaib, 1996; WHO, 1999). These advances, as well as the regular use of portable computers to collect and analyze data on-site and immediately postdisaster meant that data-driven disaster relief activities could begin more quickly than traditional methods would have allowed (Lillibridge et al., 1993).

Based on these new protocols, cluster sampling became the gold standard for RNAs globally. For example, following an earthquake in Turkey, an assessment was conducted to determine the needs of disaster-displaced individuals (Daley, Karpati, & Sheik, 2001). Using modified cluster sampling and portable computers to analyze data on-site, the assessment was completed in just 10 days after the earthquake, allowing critical concerns related to shelter, food, and hygiene to be addressed (Daley et al., 2001). Other examples of rapid assessments using cluster sampling included responses to the 1998 floods in Bangladesh (Hossain & Kolsteren, 2003), Hurricane Andrew in South Florida (Hlady et al., 1994), and the First Chechen War between the Russian Federation and the Chechen Republic in 1994–96 (Drysdale, Howarth, Powell, & Healing, 2000).

Public Health Surveillance and Tracking Systems

As disaster management became more standardized through broader implementation of incident management systems in the United States, Australia, New Zealand, Brazil, and elsewhere, the collection of standardized surveillance data became a more important part of tracking the health impacts of disasters. Several reports published in the 1990s demonstrated the effective use of surveillance data in disaster management, when it is vital to fully understand the impact of a disaster on a population's health. For example, following the Rwandan genocide of 1994, mortality and morbidity surveillance systems were established to track the health impacts among Rwandan refugees settled in Zaire (Goma Epidemiology Group, 1995). Ongoing morbidity surveillance was used to observe patterns in disease, assess needs, and effectively implement interventions. For example, the system was used to track cases of diarrheal disease, such as cholera and dysentery, which tended to be highly prevalent among these large refugee populations and could be easily transmissible in small areas. After the 1991–92 famine in Somalia, reviews of data from multiple assessments led researchers to call for the establishment of a surveillance system to track mortality, morbidity, and nutritional status so that data could be used to develop interventions programs and policies (Boss et al., 1994). The standardization of surveillance systems and the sharing of the data collected made it possible for other response groups to utilize similar approaches in other disaster settings.

Epidemiology Investigations and Studies

As highlighted previously, modifications in cluster-sampling approaches made it possible to estimate the total number of people living in the disaster area and the total population in need of assistance or disaster relief (Malilay et al., 1996). Other researchers modified the cluster sampling approach to conduct national surveys to measure multiple objectives (Turner et al., 1996). In addition, cross-sectional or prevalence studies were also beginning to be used in disaster epidemiology investigations due to their relative efficiency and cost-effectiveness. For example, Chen et al., (2015) conducted face-to-face interviews with victims of the 1998 Hunan, China floods who had been diagnosed with posttraumatic stress disorder (PTSD) to better understand risk factors for chronic PTSD. Although the results from cross-sectional studies cannot be used for causal inference since they cannot establish temporality, they can be used as a basis for developing case-control or cohort studies to identify potential causal pathways.

2000s

Rapid Needs Assessments

After 2000, the focus of RNAs shifted to achieving faster response times, shorter completion times, and the development of standardized tools to systematically collect and analyze data. Since an outside team traveling to reach a disaster-effected area to conduct an RNA can sometimes take more than a week and developing protocols to support the implementation of the assessment can take additional time, disaster-effected and displaced populations may be waiting too long to gain maximum benefit from the assessment findings. For example, during the first several weeks after a disaster, the risk factors and conditions that support the transmission of communicable diseases may begin to increase (Asari, Koido, Nakamura, Yamamoto, & Ohta, 2000). One way of shortening the length of time it takes to begin an assessment is to use geographical informational system (GIS) tools to speed up the processes of sample selection (Waring, Reynolds, D'Souza, & Arafat, 2002). Handheld devices such as tablets, smart phones, and global positioning system devices can be used to collect data electronically, reducing the time needed for data collection (CDC, 2004). Electronic data collection also eliminates double data entry, which can be an important consideration when dealing with the time constraints inherent in conducting assessments during public health emergencies (Horney, Ramsey, Smith, Johnson, & MacDonald, 2011).

Proposals for standardized minimum data sets needed to respond to the public health impacts of disasters were also developed for use by public health authorities and relief agencies (Bradt & Drummond, 2003). Differences in data collection instruments can make it difficult for institutions to combine data from multiple sources or to make policy decisions or develop programs to address the immediate needs of disaster-displaced populations, particularly in the case of major disasters. Standardized assessment tools that could be used to eliminate inconsistencies that can arise when disaster managers adopt different protocols to conduct assessments at different times and locations and in response to different types of disasters were developed. For example, after Hurricane Katrina, numerous agencies and organizations conducted needs assessments of evacuees. Among evacuees relocated to West Virginia after Hurricane Katrina, individual-level data were collected via a rapid assessment developed by the West Virginia Department of Health and Human Resources and CDC. Rapid assessment data were then linked with Red Cross household registration records by name and age to ensure the rapid procurement of services needed to meet the immediate needs of evacuees (Ridenour, Cummings, Sinclair, & Bixler, 2007).

Public Health Surveillance and Tracking Systems

In 2005, the World Health Assembly adopted new International Health Regulations that established a global surveillance system for public health emergencies of international concern (Baker & Fidler, 2006). These new regulations established a list of conditions for which a single case would constitute a public health emergency, including emerging and reemerging diseases such as severe acute respiratory syndrome (SARS) and new influenza subtypes. In addition, new biosurveillance systems were implemented that integrated electronic medical records and syndromic surveillance from emergency rooms and emergency medical services in new ways that would be useful when monitoring mortality and morbidity resulting from natural disasters (Chughtai, DeVore, Kan, & Streichert, 2016). Syndromic surveillance systems were also adapted and expanded to monitor public health during large events, such as the 2012 London Olympics (Todkill et al., 2016) and the annual Hajj in the Kingdom of Saudi Arabia (Al-Tawfiq, Gautret, Benkouiten, & Memish, 2016).

Epidemiology Investigations and Studies

Recent years have seen increases in the number and intensity of both natural and man-made disasters. Since 1980, hurricanes and tropical storms causing more than $1 billion in damages have increased from an average of 0.4 per year to more than 1 per year (National Research Council, 2014). The frequency and severity of complex humanitarian emergencies, defined as acute situations in which mortality substantially increases above the population baseline, either directly because of violence, or indirectly due to malnutrition or the transmission of communicable diseases, has also increased since the 1980s and 1990s (Salama, 2004). New areas for epidemiology investigations and studies postdisaster have also emerged, including reproductive (Horney, Williams, Hsia, & Zotti, 2012; Zotti, Williams, Robertson, Horney, & Hsia, 2013) and mental health (Galea, Nandi, & Vlahov, 2005; Neria, Galea, & Norris 2009). Changes in threats—pandemic influenza, suicide bombers—and in health systems and policies—emergency department overcrowding, the Patient Protection and Affordable Care Act—mean that the body of knowledge related to disaster epidemiology can quickly become outdated, requiring the design and implementation of new studies in response to future disasters to continue to build the knowledge base of disaster epidemiology (Noji, 2005).

References

Adams, P. R., & Adams, G. R. (1984). Mount Saint Helens's ashfall: Evidence for a disaster stress reaction. *American Psychologist, 39*(3), 252.

Al-Tawfiq, J. A., Gautret, P., Benkouiten, S., & Memish, Z. A. (2016). Mass gatherings and the spread of respiratory infections: Lessons drawn from the Hajj. *Annals of the American Thoracic Society, 13*(6), 759–765.

Asari, Y., Koido, Y., Nakamura, K., Yamamoto, Y., & Ohta, M. (2000). Analysis of medical needs on day 7 after the tsunami disaster in Papua New Guinea. *Prehospital and Disaster Medicine, 15*(02), 9–13.

Baker, M. G., & Fidler, D. P. (2006). Global public health surveillance under new international health regulations. *Emerging Infectious Diseases, 12*(7).

Bertazzi, P. A. (1989). Industrial disasters and epidemiology: A review of recent experiences. *Scandinavian Journal of Work, Environment & Health*, 85–100.

Boss, L. P., Toole, M. J., & Yip, R. (1994). Assessments of mortality, morbidity, and nutritional status in Somalia during the 1991–1992 famine: Recommendations for standardization of methods. *Jama, 272*(5), 371–376.

Bouwer, L. M., Crompton, R. P., Faust, E., Höppe, P., & Pielke, R. A., Jr. (2007). Confronting disaster losses. *Science-New York Then Washington, 318*(5851), 753.

Bradt, D. A., & Drummond, C. M. (2003). Rapid epidemiological assessment of health status in displaced populations—an evolution toward standardized minimum essential data sets. *Prehospital and Disaster Medicine, 18*(01), 178–185.

Centers for Disease Control (CDC). (1992). Rapid health needs assessment following hurricane Andrew—Florida and Louisiana, 1992. *Morbidity and Mortality Weekly Report, 41*(37), 685.

Centers for Disease Control and Prevention (CDC). (2000). Morbidity and mortality associated with hurricane Floyd—North Carolina, September—October 1999. *Morbidity and Mortality Weekly Report, 49*(17), 369.

Centers for Disease Control and Prevention (CDC). (2004). Rapid community health and needs assessments after Hurricanes Isabel and Charley—North Carolina, 2003–2004. . *Morbidity and Mortality Weekly Report, 53*(36), 840.

Chen, L., Tan, H., Cofie, R., Hu, S., Li, Y., Zhou, J., et al. (2015). Prevalence and determinants of chronic post-traumatic stress disorder after floods. *Disaster Medicine and Public Health Preparedness, 9*(5), 504–508.

Chughtai, S., DeVore, K., Kan, L., & Streichert, L. C. (2016). Assessment of local health department utility of syndromic surveillance: Results of the 2015 biosurveillance needs assessment survey. *Journal of Public Health Management and Practice, 22*, S69–S74.

Daley, W. R., Karpati, A., & Sheik, M. (2001). Needs assessment of the displaced population following the August 1999 earthquake in Turkey. *Disasters, 25*(1), 67–75.

Dhara, V. R., & Dhara, R. (2002). The Union Carbide disaster in Bhopal: A review of health effects. *Archives of Environmental Health: An International Journal, 57*(5), 391–404.

Drysdale, S., Howarth, J., Powell, V., & Healing, T. (2000). The use of cluster sampling to determine aid needs in Grozny, Chechnya in 1995. *Disasters, 24*(3), 217–227.

Erikson, K. T. (1976). *Everything in its path: Destruction of community in the Buffalo Creek flood*. New York: Simon &Schuster.

Frerichs, R. R., & Tar, K. T. (1988a). Breastfeeding, dietary intake and weight-for-age of children in rural Burma. *Asia Pacific Journal of Public Health, 2*, 16–21.

Frerichs, R. R., & Tar, K. T. (1988b). Use of rapid survey methodology to determine immunization coverage in rural Burma. *Journal of Tropical Pediatrics, 34*(3), 125–130.

Galea, S., Nandi, A., & Vlahov, D. (2005). The epidemiology of post-traumatic stress disorder after disasters. *Epidemiologic Reviews, 27*(1), 78–91.

Gall, M., Borden, K. A., Emrich, C. T., & Cutter, S. L. (2011). The unsustainable trend of natural hazard losses in the United States. *Sustainability, 3*(11), 2157–2181.

Glass, R. I., O'Hare, P., & Conrad, J. L. (1979). .Health consequences of the snow disaster in Massachusetts, February 6, 1978. *American Journal of Public Health, 69*, 1047–1049.

Goma Epidemiology Group. (1995). Public health impact of Rwandan refugee crisis: What happened in Goma, Zaire, in July, 1994? *Lancet, 345*(8946), 339–344.

Hlady, W. G., Quenemoen, L. E., Armenia-Cope, R. R., Hurt, K. J., Malilay, J., Noji, E. K., et al. (1994). Use of a modified cluster sampling method to perform rapid needs assessment after Hurricane Andrew. *Annals of Emergency Medicine, 23*(4), 719–725.

Horney, J. A., Ramsey, S., Smith, M., Johnson, M., & MacDonald, P. D. M. (2011). Implementing mobile geographic information system technology in North Carolina to enhance emergency preparedness: Evaluation of associated trainings and exercises. *Journal of Emergency Management., 9*(5), 47–55.

Horney, J. A., Williams, A., Hsia, J., & Zotti, M. (2012). Cluster sampling with referral to improve the efficiency of estimating unmet needs among pregnant and postpartum women after disasters. *Women's Health Issues: Official Publication of the Jacobs Institute of Women's Health, 22*(3), 253–257.

Hossain, S. M., & Kolsteren, P. (2003). The 1998 flood in Bangladesh: Is different targeting needed during emergencies and recovery to tackle malnutrition? *Disasters, 27*(2), 172–184.

Ivanov, V. K., Tsyb, A. F., Khait, S. E., Kashcheev, V. V., Chekin, S. Y., Maksioutov, M. A., et al. (2012). Leukemia incidence in the Russian cohort of Chernobyl emergency workers. *Radiation and Environmental Biophysics, 51*(2), 143–149.

Killian, L. M. (1954). *Evacuation of Panama city residents before hurricane florence*. Washington, DC: National Academy of Science.

Killian, L. M. (1956). *An Introduction to methodological problems of field studies in disasters* (A special report prepared for the Committee on Disaster Studies, Publication No. 465.). Washington, DC: National Academy of Sciences, National Research Council.

Korteweg, H. A., van Bokhoven, I., Yzermans, C. J., & Grievink, L. (2010). Rapid health and needs assessments after disasters: A systematic review. *BMC Public Health, 10*(1), 295.

Laska, S., & Peterson, K. (2011). The convergence of catastrophes and social change the role of participatory action research in support of the new engaged citizen. *Journal of Applied Social Science, 5*(1), 24–36.

Lemeshow, S., & Robinson, D. (1985). Surveys to measure programme coverage and impact: A review of the methodology used by the expanded programme on immunization. *World Health Statistics Quarterly, 38*, 65–75.

Lillibridge, S. R., Noji, E. K., & Burkle, F. M. (1993). Disaster assessment: The emergency health evaluation of a population affected by a disaster. *Annals of Emergency Medicine, 22*(11), 1715–1720.

Logue, J. N., Hansen, P. H., & Hansen, H. (1980). A case-control study of hypertensive women in a post-disaster community: Wyoming Valley, Pennsylvania. *Journal of Human Stress, 6*(2), 28–34.

Malilay, J., Flanders, W. D., & Brogan, D. (1996). A modified cluster-sampling method for post-disaster rapid assessment of needs. *Bulletin of the World Health Organization, 74*(4), 399.

Malilay, J., Heumann, M., Perrotta, D., Wolkin, A. F., Schnall, A. H., Podgornik, M. N., et al. (2014). The role of applied epidemiology methods in the disaster management cycle. *American Journal of Public Health, 104*(11), 2092–2102.

Mileti, D. (1987). Sociological methods and disaster research. In R. Dynes, B. de Marchi, & C. Pelanda (Eds.), *Sociology of disasters: Contributions of sociology to disaster research*. Milan, Italy: Fanco Angeli.

National Institutes of Health Disaster Research Response (DR2). (n.d.). Research planning, protocols, and funding. Retrieved from https://dr2.nlm.nih.gov/protocols.

National Research Council. (2014). *Reducing coastal risk on the East and Gulf coasts*. Washington, DC: National Academies Press.

Neria, Y., Galea, S., & Norris, F. H. (Eds.). (2009). *Mental health and disasters*. Cambridge University Press.

Nishikiori, N., Abe, T., Costa, D. G., Dharmaratne, S. D., Kunii, O., & Moji, K. (2006). Who died as a result of the tsunami?—Risk factors of mortality among internally displaced persons in Sri Lanka: A retrospective cohort analysis. *BMC Public Health, 6*(1), 1.

Noji, E. K. (2005). Disasters: Introduction and state of the art. *Epidemiologic Reviews, 27*(1), 3–8.

Ogden, C. L., Gibbs-Scharf, L. I., Kohn, M. A., & Malilay, J. (2001). Emergency health surveillance after severe flooding in Louisiana, 1995. *Prehospital and Disaster Medicine, 16*(03), 138–144.

Rahu, K., Rahu, M., Tekkel, M., & Bromet, E. (2006). Suicide risk among Chernobyl cleanup workers in Estonia still increased: An updated cohort study. *Annals of Epidemiology, 16*(12), 917–919.

Raska, K. (1966). National and international surveillance of communicable diseases. *WHO Chronicle, 20*(9), 315–321.

Ridenour, M. L., Cummings, K. J., Sinclair, J., & Bixler, D. (2007). Displacement of the underserved: medical needs of Hurricane Katrina evacuees in West Virginia. *Journal of Health Care for the Poor and Underserved, 18*(2), 369–381.

Savitz, D. A., Oxman, R. T., Metzger, K. B., Wallenstein, S., Stein, D., Moline, J. M., et al. (2008). Epidemiologic research on man-made disasters: Strategies and implications of cohort definition for World Trade Center worker and volunteer surveillance program. *Mount Sinai Journal of Medicine: A Journal of Translational and Personalized Medicine, 75*(2), 77–87.

Sommer, A., & Mosley, W. (1972). East Bengal cyclone of November, 1970: Epidemiological approach to disaster assessment. *Lancet, 299*(7759), 1030–1036.

Spencer, H. C., Romero, A., & Feldman, R. A. (1977). Disease surveillance and decision-making after the 1976 Guatemala earthquake. *Lancet, 2*, 181–184.

Stewart, P. A., Stenzel, L. M., Kwok, R. K., Branch, E., Blair, A., Engel, L. S., et al. (2011). *Background and strategy for exposure assessment for the "Gulf Long-Term Follow-Up Study"*.

Tierney, K. J. (2007). From the margins to the mainstream? Disaster research at the crossroads. *Sociology, 33*(1), 503.

Todkill, D., Hughes, H. E., Elliot, A. J., Morbey, R. A., Edeghere, O., Harcourt, S., et al. (2016). An observational study using english syndromic surveillance data collected during the 2012 London Olympics-what did syndromic surveillance show and what can we learn for future mass-gathering events? *Prehospital and Disaster Medicine, 1*.

Turner, A. G., Magnani, R. J., & Shuaib, M. (1996). A not quite as quick but much cleaner alternative to the Expanded Programme on Immunization (EPI) Cluster Survey design. *International Journal of Epidemiology, 25*(1), 198–203.

Waring, S. C., Reynolds, K. M., D'Souza, G., & Arafat, R. R. (2002). Rapid assessment of household needs in the Houston area after tropical storm Allison. *Disaster Management & Response: DMR: an Official Publication of the Emergency Nurses Association*, 3–9.

World Health Organization. (1999). *Rapid health assessment protocols for emergencies*. Geneva, Switzerland: World Health Organization.

Zotti, M., Williams, A., Robertson, M., Horney, J. A., & Hsia, J. (2013). Post-disaster reproductive health outcomes. *Maternal and Child Health Journal, 17*(5), 783–796.

CHAPTER

2

Methods: Surveillance

Kahler Stone, Jennifer A. Horney
Texas A&M University, College Station, TX, United States

THE HISTORY OF PUBLIC HEALTH SURVEILLANCE

In May of 1959, the World Health Assembly (WHA) prioritized the eradication of smallpox in the coming years and tasked the World Health Organization (WHO) with leading the effort. Smallpox is a highly contagious viral disease that has no treatment and can lead to death in its severe manifestations (Fig. 2.1). The virus is transmitted through direct contact with infected persons who have a signature rash. In 1959, there were an estimated 77,555 cases of smallpox worldwide; the total number was likely higher due to underreporting. While the eradication of smallpox had been debated for more than a decade, the idea that eradication was possible was fueled by a report to the WHA that highlighted the importance and feasibility of eradicating smallpox by a Russian virologist, Dr. Viktor Zhdanov.

The Smallpox Eradication Program (SEP) was not the first attempt at implementing an eradication program. A number of other diseases had their own programs, most notably malaria. However, the SEP, in contrast to the effort to eradicate malaria, had an effective weapon to control and eliminate smallpox—a vaccine. The vaccine, first developed by Edward Jenner in 1796, was by the 1950s stable enough to be transported around the globe in search of endemic areas. By 1977, the last naturally occurring case of smallpox was reported, leading officials to declare the eradication of smallpox in 1980.

This astounding public health achievement has been rightly attributed to the effectiveness of the vaccine and the mass coordination of its delivery. Without a vaccine, this "disgusting smallpox disease," as Thomas Jefferson called it in a letter to Edward Jenner, would still likely be present today (Fenner, Arita, Jezek, & Ladnyi, 1988). However, an often overlooked, yet critical component that enabled the program's success was the substantial public health surveillance effort that took place. As the SEP started administering vaccines and putting control measures in place, surveillance data from areas where the disease was present were critical to the strategic planning efforts of the program.

Disaster Epidemiology
http://dx.doi.org/10.1016/B978-0-12-809318-4.00002-2

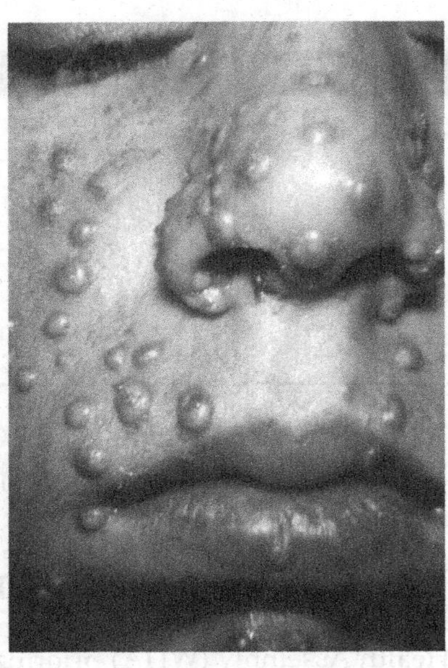

FIGURE 2.1 Maculopapular lesions on the face of a 15 year-old boy infected with smallpox. *Photo Credit: Centers for Disease Control and Prevention/J. D. Millar, MD, 1972.*

During the eradication campaign, both passive and active surveillance was used. For example, as the SEP efforts intensified during the 1960s, jurisdictions began producing total counts of reported smallpox cases by area every two to three weeks using passive surveillance. Before 1967, the main measure used to evaluate the success of SEP was the total number of vaccines administered in endemic countries. Little attention was paid to smallpox surveillance data. However, after 1967, surveillance data became the most important tool used for the strategic distribution of vaccine. Surveillance data not only concentrated vaccine efforts in areas of high incidence but also aided in the development of SEP's public education and mass media campaigns. Surveillance, along with vaccine coverage community assessments, played a key role in targeting vaccine distribution. Surveillance of smallpox also allowed for the epidemiologic analysis of demographic characteristics of the cases, improving the efficiency of the SEP focusing vaccine delivery on certain groups rather than on endemic countries as a whole.

Another key feature of smallpox surveillance was the effective dissemination of surveillance reports. Sharing data related to where smallpox was being reported to all those who were working to eradicate the disease was critical. Sharing data also allowed for the implementation of additional active surveillance in areas where outbreaks were being reported. Active surveillance during outbreaks allowed teams to vaccinate entire communities to prevent further spread of the virus. In a 1988 WHO report, key leaders of the eradication effort pointed out that it was only logical to rely heavily on public health surveillance in effort to eradicate smallpox; without surveillance there would be no way to know when the program reached zero cases (Fenner et al., 1988).

DEFINING SURVEILLANCE

Surveillance is considered an essential capability for public health. In addition to the worldwide disease control program described above, public health surveillance has been used in many different ways, including the regular collection of data during or after a disaster or other major event.

Advances in electronic data collection and implementation of new technologies for detecting diseases and adverse health events have led for some to call for updating the definition of public health surveillance. However, public health experts have maintained that current definitions continue to adequately capture the meaning and significance of public health surveillance (Hall, Correa, Yoon, & Braden, 2012). For example, WHO's International Health Regulations define surveillance as "...the systematic ongoing collection, collation and analysis of data for public health purposes and the timely dissemination of public health information for assessment and public health response as necessary" (World Health Organization, 2008). The Center for Disease Control and Prevention (CDC) defines surveillance as "the ongoing, systematic collection, analysis, and interpretation of health data, essential to the planning, implementation and evaluation of public health practice, closely integrated with the dissemination of these data to those who need to know and linked to prevention and control" (Center for Disease Control and Prevention, 2012a). The International Epidemiological Association defines surveillance as "...the systematic, ongoing collection, management, analysis, and interpretation of data followed by the dissemination of these data to public health programs to stimulate public health action" (Porta, 2008).

Each of these definitions can be distilled into several key components. First, public health surveillance involves the collection of health data from a population of interest. While data can vary in their level of specificity, they usually include specific notifiable conditions or diseases of public health concern (i.e., those that are communicable or have potential bioterror characteristics), along with case definitions for each condition or disease (Center for Disease Control and Prevention, 2016a,b; World Health Organization, 2008). This type of surveillance data can be obtained through passive reporting from health-care providers and schools or can be actively sought out in an effort to gather more extensive data about a particular condition. Whether passive or active, surveillance involves systematically obtaining regular data on the health of the population to provide complete and consistent counts of certain diseases.

The second component of public health surveillance is the analysis and interpretation of data. Descriptive epidemiology is used to characterize patterns and trends in data across person, place, and time. A baseline number of cases can be established for various conditions and diseases to inform public health practitioners when certain conditions have reached a higher than expected case count or an outbreak is potentially in progress. Furthermore, the analysis of surveillance data allows epidemiologists to explore how diseases are presenting in a population. Are there more cases among men or women? Are cases predominately among certain racial or ethnic groups? Are cases equally distributed across space or clustered in one neighborhood?

The final essential component in the definition of public health surveillance is the dissemination of data for public health action. For example, 350 laboratories throughout the country report the number of influenza positive results to the CDC weekly (Center for Disease Control and Prevention, 2016c). These data are analyzed to determine which strains of influenza are being reported and if the number of reported cases falls within the expected range for the seasonal period (e.g., more cases are expected during the winter months in the Northern Hemisphere's influenza season). The results of these analyses are compiled and routinely disseminated back to reporting health-care providers, schools, and workplaces, as well as to the general public to increase awareness of influenza activity in the community and help to promote influenza vaccination and nonpharmaceutical interventions such as handwashing.

Case definitions are a critical part of public health surveillance, providing the criteria by which we count the number of people who have a specific condition or disease. A case definition serves as the basis for deciding what is classified as a case and what is not. Epidemiologists use case definitions for routine public health surveillance, during outbreak investigations, and when irregular health events are occurring or when a condition is newly emerging or reemerging. For example, Salmonellosis is a nationally notifiable condition in the United States. There are three distinct case definitions for salmonellosis including suspected, probable, and confirmed. For salmonellosis, a suspect case meets the suspect laboratory criteria for diagnosis, while a probably case can be epidemiologically linked to a confirmed case, and a confirmed case meets the confirmed laboratory criteria for diagnosis (Center for Disease Control and Prevention, 2012b). These definitions allow epidemiologists to count only the cases of diarrheal diseases that are confirmed to have been caused by the *Salmonella* spp. pathogen as Salmonellosis, rather than counting all diarrheal diseases that have similar symptoms as Salmonellosis.

During a disaster, both active and passive surveillance can be conducted to capture morbidity and mortality data to better understand how an event is adversely affecting the populations' health. Morbidity data include those that are ill, diseased or injured. Morbidity surveillance in disaster settings is not only for infectious disease outbreaks but also for injuries and other chronic ailments the disaster may be causing, including mental illnesses (Schnall et al., 2011). Collecting incidence and prevalence data, and calculating stratified rates of morbidity after a disaster informs emergency management and public health officials' decisions for resource allocation and targeted interventions (Center for Disease Control and Prevention, 2014).

Mortality data include deaths and can be an important indicator to the extent of the disaster (Agency for Toxic Substances and Disease Registry, 2016). Mortality surveillance in a disaster is one measure of the scale of the disaster's impact on the population (Malilay et al., 2014). Mortality data can be used to estimate the number of event-related deaths and characterize them through the collection of demographic information, time and location of the death, and the cause of death, if known (Ragan, Schulte, Nelson, & Jones, 2008).

TYPES OF SURVEILLANCE

Passive Surveillance

Passive surveillance is the most common form of surveillance, mainly because it demands few resources compared with active surveillance. Several national passive surveillance systems operate in the United States. The National Notifiable Disease Surveillance System (NNDSS) relies on health-care professionals to report notifiable conditions to their local or state health department (Center for Disease Control and Prevention, 2016b). For notifiable conditions, CDC works closely with states to encourage and enforce reporting that is consistent with each state's own laws (Roush, Birkhead, Koo, Cobb, & Fleming, 1999). State health departments, and subsequently CDC, wait for disease reports to come in, rather than actively seeking cases. While the pervasiveness of electronic medical records in the United States mean that case reports and other surveillance data are sent electronically to improve timeliness and efficiency, often, health-care providers and laboratories who report conditions to health departments provide limited information about each case being reported. After receiving this information, state or local public health epidemiologists follow-up on some cases to obtain additional relevant information (e.g., demographics, occupation, residence, and potential contacts). Although the step of gathering additional information on the reported cases may seem active, it is not considered active surveillance because the health department did not seek out the case; it was reported to them from the provider. Because passive surveillance relies heavily on health-care providers and facilities to report conditions to public health agencies on their own accord, it is not uncommon for certain conditions to be underreported. For example, only 20% of shigellosis cases seek medical care, and only a fraction of those have a stool specimen collected (Scallan et al., 2006). Therefore, even if all those cases are reported to public health authorities, the number of cases will be an underestimate and passive surveillance will not have provided a good estimate of the actual burden of shigellosis in the population.

There are several ways to enhance passive surveillance. For example, health-care providers can be encouraged to report to their public health authorities through education on reporting laws and requirements, as well as penalties and public health implications associated with noncompliance. Federal, state, and local public health can provide frequent and relevant feedback on the conditions being reported to reinforce the importance of reporting and make it as simple as possible for providers to report, such as by accepting automatic electronic reporting (Jajosky & Groseclose, 2004). The usefulness of passive surveillance is directly related to the completeness of the data reported. Complete reporting can only be accomplished when public health authorities work closely with providers and actively seek out additional case information in response to reporting, including contact tracing and identification of additional possibly exposed individuals.

In disaster settings, existing passive surveillance systems often times remain intact; however, there may also be a need for expanded surveillance. For example, in 1995, the Alfred P. Murrah Federal Building in Oklahoma City was bombed, resulting in 592 physical injuries and 167 deaths (Mallonee et al., 1996). To capture the impact this disaster had on the population, officials from the Oklahoma State Department of Health (OSDH) immediately made injuries and fatalities associated with the bombing reportable to public health authorities (Table 2.1). This allowed the Oklahoma City-County Health Department (OKC-County) and OSDH to analyze surveillance data on the injuries sustained from the explosion. Due to injury reporting from health-care providers, hospitals, and clinics, OSDH and OKC-County were able to construct a complete registry of injured individual to be used for psychiatric referral and follow-up in later studies (North et al., 1999). Making the injuries after this incident notifiable alleviated the need for OSDH to establish active surveillance to

TABLE 2.1 Classifications of Injuries Among 592 Survivors of the Oklahoma City Bombing

Injury Type	Number of Injuries	Percentage (%)
Soft tissue injury	506	85.5
Fracture/dislocation	60	10.1
Sprain	150	25.3
Head injury	80	13.5
Ocular injury	59	10.0
Burn	9	1.5

Adapted from Mallonee, S., Shariat, S., Stennies, G., Waxweiler, R., Hogan, D., & Jordan, F. (1996). Physical injuries and fatalities resulting from the Oklahoma City bombing. JAMA, *276*(5), 382–387.

extract data from hospital and clinic records, saving time and resources for other necessary public health tasks.

Active Surveillance

Active surveillance requires public health authorities to proactively seek out cases of specific diseases by contacting health-care providers, schools, or laboratories, rather than waiting for providers to report when they identify a case. During active surveillance, a health-care facility or provider provides a complete case count from a specific time period under surveillance, including reports of zero cases. This is a key difference between passive and active surveillance. In passive surveillance, public health authorities accept there are no cases if no cases are reported, even though underreporting is assumed. In active surveillance, health departments are able to confirm that there are no cases when the reporting facility reports zero cases. Evaluation of reporting entities is also common when active surveillance is instituted to evaluate the timeliness and completeness of reporting. For example, if a school is undergoing an outbreak of gastroenteritis, public health authorities might proactively work with school staff to provide daily reports of defined cases. If the school staff fail to report daily, public health will stimulate timeliness of reporting by either contacting school staff by telephone or going in person to the school to obtain the daily case count, or line list. This proactive approach of seeking cases or conditions in defined populations means that active surveillance has greater sensitivity compared with passive surveillance (Center for Disease Control and Prevention, 2012a).

The Foodborne Diseases Active Surveillance Network (FoodNet) is an active surveillance system in the United States that is coordinated by the CDC (Center for Disease Control and Prevention, 2015). FoodNet personnel operate in 10 states and routinely interact with more than 650 clinical laboratories, actively identifying new cases of specific laboratory-confirmed foodborne illnesses. Audits of the reporting laboratories are conducted regularly to ensure that all known cases are reported. Additionally, FoodNet staff actively review hospital discharge data for any children who experienced hemolytic-uremic syndrome (HUS), which is caused by Shiga toxin–producing *Escherichia coli* (STEC). Even though STEC and HUS are passively reported through NNDSS, FoodNet goes a step further within these 10 states to review hospital records to ensure no cases of HUS caused by STEC went unreported. When any case of any of these conditions are identified in the FoodNet system, FoodNet staff collect additional data on each case for further epidemiologic investigation, including hospitalization status, patient status, and travel history. While this type of active surveillance system requires significant resources to proactively seek out foodborne illnesses—estimated at approximately $7 million per year—the seriousness and overall burden of foodborne diseases in the United States, particularly HUS in children, makes this type of active surveillance necessary (Henao, Jones, Vugia, & Griffin, 2015).

During a disaster, existing passive surveillance systems may be augmented with active surveillance, especially if normal surveillance activities have been interrupted due to the incident (CDC Disaster Surveillance). Active surveillance can also be used in disasters when temporary shelters, mobile clinics, or other short-term medical care centers are set up to monitor these sites for illnesses related to the disaster. This is sometimes referred to as shelter surveillance. After Hurricane Andrew made landfall on the US Gulf Coast in August 1992, Louisiana public health officials conducted active surveillance in cooperation with emergency departments (EDs), public utility personnel, and coroners to identify hurricane-related injuries, illnesses, and deaths. They found high rates of cuts, lacerations, and puncture wounds among those injured during the postimpact phase of the hurricane (McNabb, Kelso, Wilson, McFarland, & Farley, 1995). Similarly, after Hurricane Floyd made landfall in North Carolina in September 1999, active surveillance was established with 20 hospitals in flood-affected regions. Daily ED logs were collected and evaluated for illnesses and injuries associated with the storm during the disaster and recovery phases. Health authorities compared the data to the previous years' time period and found suicide attempts, dog bites, basic medical needs, dermatitis, arthropod bites, and asthma had significant increases after Floyd made landfall when compared with the previous year (Center for Disease Control and Prevention, 2000).

Other Types of Surveillance

Sentinel Surveillance

When data are not available or accessible for an entire population, sentinel surveillance can be used to provide sufficient information for public health action. Sentinel surveillance is conducted at specific sites or in specific populations and may be passive or active. Instead of reports on a specific condition provided in traditional active or passive surveillance, with sentinel surveillance only specific hospitals or providers report on the condition. Sentinel sites can be spread across the world, or located within a single region, country, or community, depending on the population of interest. Several situations make sentinel surveillance a good option, including when large surveillance systems are too costly or when collecting information on every case or condition would be too logistically complex. Sentinel surveillance systems offer a higher level of detail on diseases of interest and enable the identification of trends over time. Because sentinel surveillance uses only selected locations or populations, it cannot be used to detect rare conditions or make inferences to populations outside the sentinel catchment areas (World Health Organization, 2016).

A National Sentinel Site Surveillance System was implemented following the January 2010 earthquake in Haiti (Fig. 2.2). Sentinel surveillance was established by the Haitian Ministry of Health in coordination with national and international public health agencies. The sentinel surveillance system was designed to detect outbreaks, identify trends, and document the needs of affected populations for disaster relief. Daily telephone and electronic reports were received from 51 hospital and clinic sites in the affected areas on 25 specific health conditions. Using data collected through this system, public health authorities were able to successfully dispel local rumors about clusters of disease, allowing relief resources to be allocated appropriately (Centers for Disease Control and Prevention, 2010).

Syndromic Surveillance

Traditional public health surveillance methods, which have already been discussed, focus on diagnostic data and laboratory reports that can be difficult to obtain and can involve significant delays in reporting. These challenges led public health practitioners to develop innovative systems to collect different types of data, including preclinical and prediagnostic data from a variety of sources (Buckeridge et al., 2002). While diagnostic data provide traditional case identification and counts of certain diseases, this information is not typically gathered and reported in real time. For outbreak detection and public health surveillance of natural

FIGURE 2.2 Destruction of the Hotel D'Haiti in Port-au-Prince, Haiti following a magnitude-7 earthquake on January 12, 2010. *Phot Credit: U.S. Geologic Survey/Walter Mooney, 2010.*

and man-made or technological disasters, the timeliness of information is directly correlated to its relevance. For example, using preclinical and prediagnostic data can provide situational awareness for certain syndromes or population behaviors. Preclinical data can include data from emergency rooms, poison control centers, emergency medical services, pharmacies, and work or school absenteeism. In the case of emergency room data, syndromic data are collected before a health-care provider diagnoses a patient, and is based on the patients' self-reported clinical symptoms or syndromes, such as influenza-like illness (ILI) or gastrointestinal illness. The collection of close to real-time data on syndrome categories is the basis for syndromic surveillance, which allows potential aberrations in populations to be detected more quickly than a traditional outbreak could be detected by diagnostic means. Syndromic surveillance systems can also inform public health authorities of the potential magnitude of an already identified outbreak by monitoring factors such as emergency room chief complaints and absenteeism.

Specific diseases or groups of diseases typically have common signs and symptoms. When using prediagnostic or syndromic data, there are many possible self-reported complaints by patients, as well as clinical signs and symptoms, which are recorded in health-care facilities before a diagnosis is determined (e.g., fever, cough, rash). These self-reported complaints and early clinical signs and symptoms can be categorized into syndromes that may be of public health concern. For example, a common syndrome that is monitored is upper respiratory illness or ILI. One way to look for this syndrome is to look at ED visits and flag all visits with complaint records that contain key words such as, "flu," "flulike," "influenza," "fever plus cough," or "fever plus sore throat" (Wu & Kelsey, 2013). Detecting an ILI syndrome in this manner could provide an early warning of a potential widespread influenza outbreak or an intentional act of bioterrorism in a community that could precede laboratory results, giving public health officials a head start of their response.

Syndromic surveillance was first used in the mid-1990s when health departments began looking for alternative methods to detect large outbreaks. In 1995, the New York City Department of Health and Mental Hygiene created a syndromic surveillance system to detect outbreaks of diarrheal disease in nursing homes by collecting data on gastrointestinal syndromes in nursing home populations, along with laboratory reports and the sale of over-the-counter (OTC) medications in pharmacies (Heffernan et al., 2004). Following the anthrax attacks of 2001, bioterrorism preparedness became a priority for the CDC and other federal agencies, placing an increased emphasis on the development of syndromic surveillance systems to detect, evaluate, and report suspicious events (Chen, Zeng, & Yan, 2010, chap. 2). During the 2000s, much of the funding and development of syndromic surveillance systems was tailored to potential bioterror disease identification. Although syndromic surveillance can still serve those purposes today, it has largely been transformed to include identification of general disease trends in communities as longitudinal data on particular syndromes has been collected over time and across multiple reporting facilities.

Syndromic surveillance systems differ from active surveillance in that they collect data on symptoms through automated data feeds in real time or close to real time. In active surveillance, reporting facilities and public health agencies have designated staff who routinely seek out cases of specific diseases, which requires more human and fiscal resources than other types of surveillance. Syndromic surveillance, with its reliance on automatic data feeds or routine electronic queries, means that a number of signs and symptoms can be flagged across different reporting facilities and communities. The automation of syndromic surveillance allows for almost instantaneous reporting from EDs and other data sources, allowing for early detection and warning of potential disease threats in the community.

As in other types of surveillance, syndromic surveillance does not stop at data collection and alert notifications. Syndromic surveillance systems have public health responses and actions built into specific alerts. For example, monitoring purchases of OTC medications for increases in antidiarrheals may alert public health authorities to be possibility of an outbreak of a waterborne disease such as cryptosporidiosis or cyclosporiasis and allow public health officials to start an investigation. This type of early warning also allows public health officials to begin notifying local health-care providers of the potential for a waterborne disease outbreak and to encourage laboratory testing for specific pathogens, since waterborne diseases are generally underreported and not routinely tested for. An early signal from syndromic surveillance may allow public health agencies to start their response days or even weeks before traditional surveillance would see confirmed diagnoses being reported.

Of course, syndromic surveillance has limitations. It should be considered an additional surveillance mechanism, not a primary means of outbreak or disease detection. While prediagnostic data can be collected more quickly, they often lack specificity since reporting is done by syndrome. When public health resources are limited, such as after a disaster or in a smaller community with fewer epidemiologists, monitoring a syndromic surveillance system for alerts can be time-consuming. Effective syndromic surveillance systems are also expensive, require time to set up, and depend on strong information technology support. However, if a syndromic surveillance system is already in place (e.g., ED visits with syndrome signaling, OTC monitoring, work or school absenteeism) and a disaster occurs, it can play a role in detecting disease occurrence during the post-disaster recovery period.

USING PUBLIC HEALTH SURVEILLANCE DATA DURING A DISASTER

In 2007, New South Wales, Australia, experienced two days of severe storms in the coastal region of Newcastle (Hope et al., 2008). These storms produced an immense amount of rain and wind and resulted in both power outages and severe flooding. More than 200,000 homes were without power for at least a week and seven evacuation centers were open for displaced residents. The storm stranded a coal ship on the city's coastline, a potential environmental crisis if the ship's on-board fuel should begin leaking. During the disaster recovery, Hunter New England Population Health (HNEPH) was the lead health agency responsible for public health services, including enhanced surveillance of health events. Knowing that traditional surveillance had significant delays in detecting outbreaks with laboratory confirmations and diagnoses, HNEPH decided to utilize the existing New South Wales ED syndromic surveillance system and expand it to the area affected by the storm, which was not connected into the system. Within one week after the storms, 11 EDs from the affected area were linked to the system and generating reports four times a day. Before the EDs were brought online, active surveillance for gastroenteritis was conducted through manual extraction and submitted using the International Classification of Diseases Version 10 (ICD-10) codes. During active surveillance, hospital and public health staff expended tremendous time in the abstraction and analysis of data looking for trends that might signal an increase in gastroenteritis. Once the syndromic surveillance system data feeds were operating successfully, HNEPH were able to monitor the EDs in real time for different illness presentations, allowing them to prioritize and plan recovery efforts. No gastroenteritis outbreaks were identified during the recovery phase using these two types of surveillance, but increases in respiratory illness were detected. However, when these increases were compared with the last 5 years of data from the syndromic surveillance system, the counts of respiratory illness were not above the seasonal norm, providing decision-makers the information needed to interpret those increases as unrelated to the disaster.

CONCLUSIONS

Public health surveillance is most useful when it can be used to inform and facilitate public health action. Traditional public health surveillance systems include passive and active surveillance. Passive surveillance is the most common type of surveillance, requiring minimal resources since cases of disease are not sought out by public health authorities. Active surveillance includes public health authorities seeking specific conditions in specific areas, which is resource intensive. Sentinel surveillance is used when gathering information passively or actively from all potential reporting sources is not feasible, or when the prevalence of a condition of interest is high, leaving a high reporting burden on health providers. In sentinel surveillance, only certain locations or sites are selected to report on specified conditions, allowing for reporting and outbreak detection on trends. Syndromic surveillance utilizes automated reporting of prediagnostic data on clinical syndromes, which can be used to supplement traditional surveillance methods. Each type of surveillance has utility in disaster epidemiology, depending on the goals of public health actions.

References

Agency for Toxic Substances and Disease Registry. (2016). *Glossary of terms*. Retrieved from https://www.atsdr.cdc.gov/glossary.html.

Buckeridge, D. L., Graham, J., O'Connor, M. J., Choy, M. K., Tu, S. W., & Musen, M. A. (2002). Knowledge-based bioterrorism surveillance. In *Proceedings/annual symposium. AMIA Symposium* (pp. 76–80).

Centers for Disease Control and Prevention. (2000). Morbidity and mortality associated with hurricane Floyd–North Carolina, September–October 1999. *Morbidity and Mortality Weekly Report, 49*(17), 369–372.

Centers for Disease Control and Prevention. (2010). Launching a national surveillance system after an earthquake—Haiti, 2010. *Morbidity and Mortality Weekly Report, 59*(30), 933–938.

Centers for Disease Control and Prevention. (2012a). Principles of epidemiology in public health practice. In *An introduction to applied epidemiology and biostatistics*. U.S. Department of Health and Human Services. Retrieved from http://www.cdc.gov/ophss/csels/dsepd/ss1978/ss1978.pdf.

Centers for Disease Control and Prevention. (2012b). *Salmonellosis, 2012 case definition*. Retrieved from https://wwwn.cdc.gov/nndss/conditions/salmonellosis/case-definition/2012/.

Centers for Disease Control and Prevention. (2014). *Disaster epidemiology – frequently asked questions*. Retrieved from http://www.cdc.gov/nceh/hsb/disaster/faqs.htm.

Centers for Disease Control and Prevention. (2015). *Active laboratory surveillance*. Retrieved from http://www.cdc.gov/foodnet/surveillance.html.

Centers for Disease Control and Prevention. (2016a). *2016 nationally notifiable conditions*. Retrieved from https://wwwn.cdc.gov/nndss/conditions/notifiable/2016/.

Centers for Disease Control and Prevention. (2016b). *National notifiable disease surveillance system*. Retrieved from https://wwwn.cdc.gov/nndss/.

Centers for Disease Control and Prevention. (2016c). *Overview of influenza surveillance in the United States, seasonal influenza (flu)*. Retrieved from http://www.cdc.gov/flu/weekly/overview.htm.

Center for Disease Control and Prevention (2017). Public health surveillance during a disaster. Retrieved from https://www.cdc.gov/nceh/hsb/disaster/surveillance.htm.

Chen, H., Zeng, D., & Yan, P. (2010). *Infectious disease informatics: syndromic surveillance for public health and bio-defense*. Springer Science & Business Media.

Fenner, F., Arita, I., Jezek, Z., & Ladnyi, I. D. (1988). *Smallpox and its eradication*. Geneva: World Health Organization. Retrieved from http://www.who.int/iris/handle/10665/39485.

Hall, I., Correa, A., Yoon, P., & Braden, C. (2012). Lexicon, definitions, and conceptual framework for public health surveillance. *Morbidity and Mortality Weekly Report, 61*(3), 10–14.

Heffernan, R., Mostashari, F., Das, D., Karpati, A., Kulldorff, M., & Weiss, D. (2004). Syndromic surveillance in public health practice, New York city. *Emerging Infectious Diseases, 10*(5), 858–864. http://dx.doi.org/10.3201/eid1005.030646.

Henao, O. L., Jones, T. F., Vugia, D. J., & Griffin, P. M. (2015). Foodborne diseases active surveillance network—2 decades of achievements, 1996–2015. *Emerging Infectious Diseases, 21*(9), 1529–1536. http://dx.doi.org/10.3201/eid2109.150581.

Hope, K., Merritt, T., Eastwood, K., Main, K., Durrheim, D. N., Muscatello, D., et al. (2008). The public health value of emergency department syndromic surveillance following a natural disaster. *Communicable Diseases Intelligence Quarterly Report, 32*(1), 92–94.

Jajosky, R. A., & Groseclose, S. L. (2004). Evaluation of reporting timeliness of public health surveillance systems for infectious diseases. *BMC Public Health, 4*(29). http://dx.doi.org/10.1186/1471-2458-4-29.

Malilay, J., Heumann, M., Perrotta, D., Wolkin, A. F., Schnall, A. H., Podgornik, M. N., et al. (2014). The role of applied epidemiology methods in the disaster management cycle. *American Journal of Public Health, 104*(11), 2092–2102. http://dx.doi.org/10.2105/AJPH.2014.302010.

Mallonee, S., Shariat, S., Stennies, G., Waxweiler, R., Hogan, D., & Jordan, F. (1996). Physical injuries and fatalities resulting from the Oklahoma City bombing. *JAMA, 276*(5), 382–387.

McNabb, S. J., Kelso, K. Y., Wilson, S. A., McFarland, L., & Farley, T. A. (1995). Hurricane andrew-related injuries and illnesses, Louisiana, 1992. *Southern Medical Journal, 88*(6), 615–618.

North, C. S., Nixon, S. J., Shariat, S., Mallonee, S., McMillen, J. C., Spitznagel, E. L., et al. (1999). Psychiatric disorders among survivors of the Oklahoma City bombing. *JAMA, 282*(8), 755–762.

Porta, M. (Ed.). (2008). *A Dictionary of epidemiology* (5th ed.). Oxford; New York: Oxford University Press.

Ragan, P., Schulte, J., Nelson, S. J., & Jones, K. T. (2008). Mortality surveillance: 2004 to 2005 Florida hurricane-related deaths. *The American Journal of Forensic Medicine and Pathology, 29*(2), 148–153. http://dx.doi.org/10.1097/PAF.0b013e318175dd5e.

Roush, S., Birkhead, G., Koo, D., Cobb, A., & Fleming, D. (1999). Mandatory reporting of diseases and conditions by health care professionals and laboratories. *JAMA, 282*(2), 164–170.

Scallan, E., Jones, T. F., Cronquist, A., Thomas, S., Frenzen, P., & Hoefer, D. (2006). Factors associated with seeking medical care and submitting a stool sample in estimating the burden of foodborne illness. *Foodborne Pathogens and Disease, 3*(4), 432–438. http://dx.doi.org/10.1089/fpd.2006.3.432.

Schnall, A. H., Wolkin, A. F., Noe, R., Hausman, L. B., Wiersma, P., Soetebier, K., et al. (2011). Evaluation of a standardized morbidity surveillance form for use during disasters caused by natural hazards. *Prehospital and Disaster Medicine, 26*(2), 90–98. http://dx.doi.org/10.1017/S1049023X11000112.

World Health Organization. (2008). *International health regulations (2005)* (2nd ed.). World Health Organization.

World Health Organization. (2016). *Sentinel surveillance*. Retrieved from http://www.who.int/immunization/monitoring_surveillance/burden/vpd/surveillance_type/sentinel/en/.

Wu, F., & Kelsey, A. (2013). Early detection of influenza activity using syndromic surveillance in Missouri. *Online Journal of Public Health Informatics, 5*(1).

Vignette: Veterans Health Affairs, Veterans, and Disasters

Tiffany Radcliff[1,3], *Aram Dobalian*[2,3], *Karen Chu*[3]

[1]Texas A&M University, College Station, TX, United States; [2]University of California, Los Angeles, CA, United States; [3]VA Greater Los Angeles Healthcare System, Sepulveda, CA, United States

INTRODUCTION

The Department of Veterans Affairs (VA) operates the largest health-care system in the United States through the Veterans Health Administration (VHA). Under routine circumstances, the VHA operates 1221 outpatient clinics, 144 medical centers (VAMCs), and 132 Community Living Centers to serve approximately 9 million enrolled veterans (Veterans Affairs Data: Quickfacts, n.d.; VA Office of Geriatrics, n.d.). The VHA treats approximately 70% of this enrolled population annually, providing almost 700,000 inpatient admissions and 80 million outpatient visits across the United States. Veterans are older, more likely to be male, and less likely to be uninsured compared with nonveterans (Profile of Veterans, 2014, n.d.). Most patients that access VHA facilities for health-care services have at least one chronic health condition that requires routine medical care or monitoring. Veterans who use VHA services also have higher disability ratings, more medical conditions, and are older than other veterans, indicating a potentially more-vulnerable population (Aday, 1994; Agha, Lofgren, VanRuiswyk, & Layde, 2000; Wolinsky, Coe, Mosely, & Homan, 1985).

VETERANS AFFAIRS' ROLE IN EMERGENCY PREPAREDNESS AND RESPONSE

The VA has a core mission to provide fundamental support for disaster preparedness and response. By statute, the VA provides health care and support to veterans and communities when local emergency needs arise or when the federal government declares a disaster.

Disaster Epidemiology
http://dx.doi.org/10.1016/B978-0-12-809318-4.00003-4

As noted by Dobalian and colleagues (Dobalian, Callis, & Davey, 2001), "the extensive resources of VA as a nationwide, integrated delivery system may be used to support other federal and state agencies and local communities by providing public health and medical services after emergencies and disasters." To meet these needs, the VA has developed and maintained system capabilities within each Federal Emergency Management Agency (FEMA) region to quickly respond to a variety of disaster events with personnel, medications, supplies, facilities, and other support to ensure continuity of critical missions (Der-Martirosian et al., 2017). Examples of other resources deployed by the VA for emergencies include command and control vehicles; mobile units for housing, hygiene, pharmacy, and clinics; and dual- and multiuse vehicles for the evacuation of patients. The VA manages both its own operations and the majority of federal coordinating centers that are charged with patient care during emergencies. In addition to these responsibilities, the VA houses, rotates, and maintains a strategic national stockpile of pharmaceuticals and medical supplies to distribute, if needed, in times of disaster. As a result, the VA has a substantial investment in preparedness for emergency situations that trigger a VA response (Dobalian et al., 2001).

VETERANS HEALTH ADMINISTRATION DISASTER RESPONSES IN RECENT YEARS

Between 2005 and 2011, five VAMCs were evacuated because of extreme weather events and two of these facilities—New Orleans, Louisiana and Gulfport, Mississippi—were permanently closed due to storm damage from Hurricanes Katrina and Rita. From a patient-care perspective, emergencies can force the relocation of VA inpatients and nursing home residents, cause a sudden surge in the demand for emergency medical services, and interrupt the continuity of care for others in the affected areas.

Superstorm Sandy's Impact of Veterans Affairs Services

When Superstorm Sandy made landfall in October 2012, the VA had substantial resources invested to minimize the impact on veterans. The VA used mainstream media and webpage updates to assure timely communications regarding operations for patients, employees, and the public (see http://www.nyharbor.va.gov/emergency/index.asp for an example). The VA Office of Inspector General offered statements related to the VHA's Superstorm Sandy response:

> When Superstorm Sandy headed towards the Manhattan campus in October 2012, the facility's leadership proactively ensured the safe evacuation of 127 inpatients and all employees and arranged continued care for 20,000 outpatients at nearby VA facilities and community-based outpatient clinics. Located in the flood zone just two blocks from the East River, the facility sustained catastrophic damage. Its utilities, fire suppression system, elevators, mechanical and electrical systems, primary care clinics, and MRI machine were severely damaged. After the water receded, a "Stay Team" provided emergency safety and protective services while emergency triage staff redirected patients to health care services at locations other than the damaged Manhattan campus. In addition, mobile examination vans were activated to provide basic services, such as

vaccinations, blood-pressure checks, and laboratory work. Reintegration of patients and staff was accomplished through an organized and phased process with attention to safety of life, major equipment, and building structures. Full reintegration was achieved in May 2013. (2014, p. 2)

To illustrate the potential impacts on VHA service delivery around major disasters such as Superstorm Sandy, we examined VA administrative records for the Northeastern United States, which are identified as Veterans Integrated Service Networks 2, 4, and 5 for the 12 months before and after Sandy's landfall. We determined that the number of unique veterans who accessed VHA services in the impacted areas increased by approximately 4400 in the 12 months after the storm, with no difference in the average, median, or modal number of encounters per enrolled veteran or the rate of primary care encounters per veteran. However, some of the types of ambulatory care appeared to differ after Superstorm Sandy. During the 12 months after Superstorm Sandy, there was a decrease in laboratory services, primary care visits, X-rays, physical therapy, and emergency department visits, but an increase in the count of telephone primary-care visits and individual sessions for mental health services, and general internal medicine visits. There was no meaningful difference before versus after the storm in counts of group sessions for substance use disorders or optometry visits (see Fig. 1).

Inpatient hospital admissions fell slightly in the impacted areas during the year after Superstorm Sandy, but the three most common reasons for VA hospital stays (chest pain, pneumonia, and depression) did not change in the 12 months after the storm (see Fig. 2). There were less frequent admissions at impacted hospitals for substance abuse, congestive heart failure, posttraumatic stress disorder, and syncope during the 12 months after Superstorm Sandy.

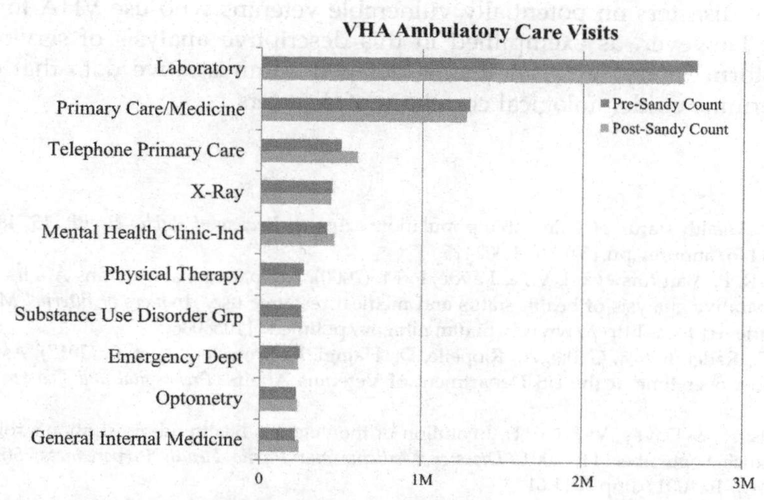

FIGURE 1 Veterans Health Administration (VHA) ambulatory care visits before and after Superstorm Sandy. Counts of the 10 most frequent reasons for ambulatory care visits in VHA clinics impacted by Superstorm Sandy for 12 months before and after the storm.

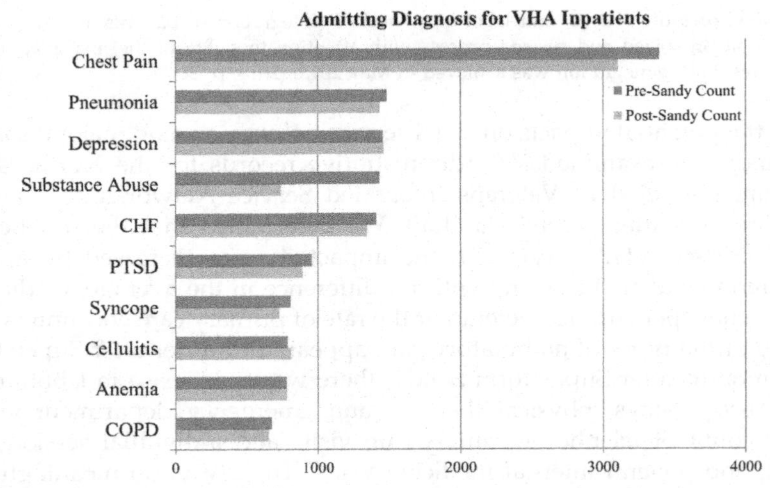

FIGURE 2 Admitting diagnosis for Veterans Health Administration (VHA) inpatients before and after Superstorm Sandy. Counts of the 10 most frequent reasons for inpatient admissions in VHA medical centers impacted by Superstorm Sandy for 12 months before and after the storm.

SUMMARY AND CONCLUSIONS

The VHA has substantial human and equipment resources to assure continuity of services for veterans and their communities in times of emergencies. More research to assess health consequences of disasters on potentially vulnerable veterans who use VHA for health care is still needed. However, as exemplified in this descriptive analysis of service utilization around Superstorm Sandy, VHA maintains detailed administrative data that can be used to quickly determine epidemiological correlates of disasters.

References

Aday, L. A. (1994). Health status of vulnerable populations. *Annual Review of Public Health, 15*, 487–509. http://dx.doi.org/10.1146/annurev.pu.15.050194.002415.

Agha, Z., Lofgren, R. P., VanRuiswyk, J. V., & Layde, P. M. (2000). Are patients at Veterans Affairs medical centers sicker? A comparative analysis of health status and medical resource use. *Archives of Internal Medicine, 160*(21), 3252–3257. Retrieved from http://www.ncbi.nlm.nih.gov/pubmed/11088086.

Der-Martirosian, C., Radcliff, T. a, Gable, A., Riopelle, D., Hagigi, F., Brewster, P., et al. (2017). Assessing hospital disaster readiness over time at the US Department of Veterans Affairs. *Prehospital and Diaster Medicine, 32*(1), 46–57.

Dobalian, A., Callis, R., & Davey, V. J. (2011). Evolution of the veterans health administration's role in emergency management since September 11, 2001. *Disaster Medicine and Public Health Preparedness, 5*(S2), S182–S184. http://dx.doi.org/10.1001/dmp.2011.61.

Geriatrics, V. O. of. (n.d.). VA Community Living Centers. Retrieved from http://www.va.gov/GERIATRICS/Guide/LongTermCare/VA_Community_Living_Centers.asp#.

Profile of Veterans, 2014. (n.d.). Retrieved from (http://www.va.gov/vetdata/docs/SpecialReports/Profile_of_Veterans_2014.pdf.

Veterans Affairs Data: Quickfacts. (n.d.). Retrieved from http://www.va.gov/vetdata/docs/Quickfacts/Homepage_slideshow_06_04_16.pdf.

Wolinsky, F. D., Coe, R. M., Mosely, R. R., & Homan, S. M. (1985). Veterans' and nonveterans' use of health services. A comparative analysis. *Medical Care, 23*(12), 1358–1371. Retrieved from http://www.ncbi.nlm.nih.gov/pubmed/3910973.

Further Reading

U.S. Department of Veterans Affairs Office of Inspector General. (2014). *Combined assessment program review of the VA New York Harbor Healthcare System, New York.* New York (Report No. 14-01293-243). Washington, DC: U.S. Department of Veteran Affairs.

CHAPTER

3

Applications: Using Information Systems to Improve Surveillance During Disasters

Laura Edison[1,2], *Karl Soetebier*[1], *Hope Dishman*[1]

[1]Georgia Department of Public Health, Atlanta, GA, United States; [2]Centers for Disease Control and Prevention, Atlanta, GA, United States

BACKGROUND

A key intent of modern information systems is to drive decision-making and action through rapid sharing and effective use of data and information among geographically disparate groups of stakeholders. The capability to do this well has become essential to public health disease surveillance and disaster response. There are three principal ingredients that must be present to ensure the successful application of information systems.

1. Networks: To share data and information, people and devices must be interconnected with each other, enabling human–human, human–machine, and machine-to-machine interoperability. The various networks that support this interconnectivity are a key consideration when preparing for and responding to disasters.
2. Flexible and adaptable information system tools: Primarily composed of Web, desktop, and mobile software applications, flexible and adaptable information system tools take advantage of interconnectivity and provide users with the capacity to access, analyze, and share surveillance data.
3. Human resources: Include developers, end users, decision-makers, and other stakeholders who capture, manage, analyze, use, and evaluate data and information system. Knowledgeable personnel with adequate training and skills are key to unlocking the full capability of these flexible tools.

During a public health emergency response, information systems can greatly improve the efficiency and effectiveness of a response, decreasing the human resources needed to achieve surveillance goals, and facilitating the collection of accurate data and rapid information

Disaster Epidemiology
http://dx.doi.org/10.1016/B978-0-12-809318-4.00004-6

sharing to support decision-making, resource allocation, and situational awareness. The use of information systems varies according to the phase of the response. For example, during the preparedness phase they can be used to perform surveillance for threats and track resources including responders. During the response phase, they can be used to perform surveillance to track morbidity and mortality, identify and track resources and responders, assess needs of the response and affected persons, and inform stakeholders and decision-makers. During the recovery phase, they can be used for ongoing surveillance of exposed persons, including responders, and for evaluating the response and the impact of the disaster.

A vast array of information system tools are used for routine public health or disease surveillance and can augment surveillance during an emergency response. Some examples include: notifiable disease reporting systems and flexible outbreak management systems that can be rapidly customized to respond to emerging threats; electronic laboratory reporting (ELR) for rapid surveillance of confirmed cases during an outbreak; syndromic surveillance systems for near real-time information about trends of customizable syndromes or overall changes in health-care burden (Atrubin, Wiese, Snider, Workman, & Mcdougle, 2013; Ayala et al., 2016; Lombardo et al., 2008); electronic surveys to create databases and survey case patients or other stakeholders during a response; free software, such as CDC's Epi Info (www.cdc.gov/epiinfo), to rapidly create databases, surveys, or data entry forms, and analyze data; and electronic death registration systems (www.naphsis.org/systems/#EDRS) to track disaster-related mortality (Baker, 2015; Howland et al., 2015; Rodgers, Paulson, Fowler, & Duffy, 2015).

In this chapter, we explore how the Georgia Department of Public Health (GDPH) has developed a flexible information system that, through the combined use of well-connected networks, flexible design, and human resource support, has successfully been used to improve surveillance and meet emergent public health needs during diverse emergency responses.

DEVELOPMENT AND USE OF THE GEORGIA STATE ELECTRONIC NOTIFIABLE DISEASE SURVEILLANCE SYSTEM

GDPH has broad information system capacity through the State Electronic Notifiable Disease Surveillance System (SendSS), an in-house developed and supported Web-based software system, which was first deployed in 1999 and has been expanding ever since. SendSS integrates a variety of networks, receives data from numerous sources, and provides a wide range of application modules and tools to support public health surveillance activities.

Networks

SendSS, built over the Oracle database platform (www.oracle.com), includes a fully integrated messaging module that orchestrates and automates the movement of data between its many trading partners. SendSS secures data transport using CDC's Public Health Information Network Messaging System (www.cdc.gov/phin/tools/phinms) software and interoperates with and manages Orion Health's Rhapsody Integration Engine (https://orionhealth.com/us/products/rhapsody/) to leverage its routing and message-handling capabilities. SendSS interfaces with the data systems of more than 130 data trading partners (hospitals, laboratories, GDPH Office of Vital Records, etc.), processes approximately 1.5 million inbound and outbound messages a month, and provides Web-enabled, internet-connected applications for the direct entry, management, and analysis of notifiable disease conditions and other public health surveillance data.

Information System Tools

SendSS consists of 39 internet application modules that support eight different program areas across GDPH. Modules and tools commonly used in emergency responses include: (1) the Outbreak Management System (OMS) that enables collection and sharing of near real-time outbreak data, line lists, laboratory test results, control measures, and contact tracing activities; (2) ELR and notifiable disease reports used to identify cases and create line lists during outbreaks; (3) syndromic surveillance, which electronically integrates and presents data from a variety of statewide sources (hospital emergency departments, acute care clinics, school nurse clinics, and influenza-like illness from sentinel providers); and (4) the survey module that allows users to create custom data capture forms that are Web-deployable and accessible with a web link. Each system module is coded within the SendSS application framework and is developed, maintained, and extended by GDPH development resources.

Human Resources

SendSS is funded by numerous GDPH program areas, and through that funding it has developed an information systems team that is housed within GDPH. These developers work closely with the program areas they support, which enables relationship building. Information systems staff understand the functions and needs of the program areas, and program staff understand the technical capabilities and limitations of the system. These close working relationships have been key to creation of user-friendly functional application modules and to their rapid adaptation when the need arises. In addition, SendSS can finely customize permissions to allow users controlled access to functions and data within modules. SendSS serves over 3700 active users including GDPH District Health Offices, County Health Departments, hospitals, private physicians, and laboratories. Disease reports, syndromic surveillance, and OMS are accessible to both state and district epidemiologists, allowing easy communication and rapid sharing of information.

Hurricane Responses

The SendSS system has provided information solutions during many emergency responses and was deployed during the GDPH 2005 Hurricane Katrina and 2016 Hurricane Matthew responses. These destructive storms displaced thousands of people in their path, and Georgia received a significant number of evacuees looking for shelter and assistance. During the Katrina response, the SendSS survey tool was used to support three response functions: (1) shelter surveillance data were collected and centrally monitored by creating an electronic SendSS survey that district epidemiologists used in the field to enter daily aggregate morbidity shelter surveillance reports. State epidemiologists then reviewed these data across multiple days to identify changing trends in illness or emerging medical needs. These reports contained information about the number and type of health-related client visits, the disposition of the client, and this information was used to facilitate additional resources or infection control education when needed in shelters; (2) hundreds of evacuees were linked to medical care by creating a SendSS survey, which was sent via email to over 25,000 Georgia physicians requesting them to volunteer their professional services to assist evacuees. The survey was used as an online registration of physician volunteers and received more than 300 responses and collected information on their location, willingness to volunteer, specialty, and availability. The Georgia Poison Center coordinated a phone bank to handle inbound requests from evacuees for medical assistance, and the SendSS survey management tool was used to search through the physician responses, matching provider volunteers with evacuees based on geography and specialty; and (3) storm-related mortality surveillance was enhanced by the creation of a multistate mortality registry used by CDC and several affected states.

During the 2016 Hurricane Matthew response, the shelter surveillance module was again used to monitor the heath of more than 2700 residents of coastal Georgia who were evacuated to shelters; clusters of respiratory and gastrointestinal illness were identified, leading to increased infection prevention education at these shelters. The survey tool was also used to rapidly create a responder safety tracking tool. Responders from across GDPH (Nursing, Environmental Health, Epidemiology, Immunizations, Emergency Preparedness, Emergency Medical Services, etc.) were asked to register themselves using a Web-based SendSS survey. This survey asked demographic information, work information, and allowed responders to self-report if responders had health concerns that may affect them during deployment. Once they were registered, they received a daily email with a link to another survey in which they could document their location, duty hours, and any exposure of injuries they encountered while deployed. The responder safety management tool was used by epidemiologists to monitor 130 GDPH responders; seven responders reported exposures that may pose a health risk or injuries and were promptly contacted to determine whether they had ongoing needs or required monitoring for disease exposures. Syndromic surveillance was also monitored during this response and identified a cluster of persons presenting to emergency departments resulting from carbon monoxide exposure from generators used because of the widespread power outages; this finding lead to public messaging about the dangers and safe use of generators.

H1N1 Response

Many SendSS components were used during the 2009–10 H1N1 influenza pandemic response in Georgia, including syndromic surveillance, ELR, and OMS. Prior to the H1N1 response, ELR data were not yet included in the data used for routine influenza surveillance. The SendSS team worked with data received from the three laboratories testing more than 90% of Georgia H1N1 cases and developed a mechanism for processing these messages, which were unique to each laboratory. The SendSS team mapped the data received in the messages to the electronic case report form created using the survey feature in OMS. Data were displayed in the OMS line list and were searchable based on case definition, geographic location, and other variables within the line list. Having key personnel in place enabled rapid system development necessary to create this interconnection that automatically processed over 15,000 H1N1 results, freeing epidemiology resources to concentrate on other aspects of managing the pandemic. Calls to the technical team to make modifications occurred frequently throughout the lengthy response effort, and the system was adapted in steps as the epidemiologic guidance and response were changed.

The syndromic surveillance module was used to examine trends of influenza-like illness; the system's capabilities include tabular, graphical, and spatial representations of patient visits to Georgia emergency departments. Georgia Flu Trends, a visualization tool built specifically for influenza-like illness, was used to combine weekly visits identified in syndromic surveillance across multiple flu seasons with map-based visualizations and influenza viral isolate data from ELRs (Fig. 3.1). The existing capabilities of the SendSS system, modified slightly to include H1N1 specific data, greatly informed DPH's understanding of how influenza was impacting the health of Georgians during the pandemic.

The state health director asked the SendSS team to develop a school absentee reporting system to better understand the impact of the pandemic on Georgia school systems using absenteeism as a proxy measure for the impact of H1N1 on the communities the school systems serve. The team developed and deployed this new module in less than two weeks, and it was made available to 178 school districts, providing information about absenteeism, which became a useful tool for situational awareness. The survey tool was used again postpandemic to perform an assessment of the impact of CDC's nonpharmaceutical interventions and to assess how prepared Georgia schools were for dealing with the consequences of the pandemic (Nasrullah et al., 2012).

2014 Ebola Virus Disease Response

One last example of the importance of Web-accessible, flexible systems and the personnel to modify them to support disaster response is found in GDPH's use of SendSS to support active monitoring of travelers during the 2014–15 Ebola virus disease (EVD) outbreak. In the fall of 2014, it was determined that travelers entering the United States with recent travel to EVD-affected countries should be monitored on their return to the United States to rapidly identify, isolate, and treat ill travelers and prevent the spread of EVD. GDPH learned that there were a large number of travelers returning to Georgia, and that the start of active monitoring was only a week away. With limited human resources to conduct this monitoring, epidemiologists turned to SendSS. The SendSS team, in close cooperation with epidemiologists, developed an online Ebola Active Monitoring System (EAMS) within SendSS to assist in the management of the two daily symptom checks and the collection of various risk information associated with each traveler. EAMS was built and deployed in 5 days and then modified over the following months as the need arose.

EAMS contained four components: (1) a traveler component that enabled the traveler to record daily symptoms online using a computer or smartphone; (2) a public health component that allowed GDPH epidemiologists to manage travelers throughout their active monitoring period and immediately notified epidemiologists by email when a traveler reported having symptoms; (3) a reporting component that provided summary statistics, a line list of travelers, and a summary report to assist with weekly reporting to CDC

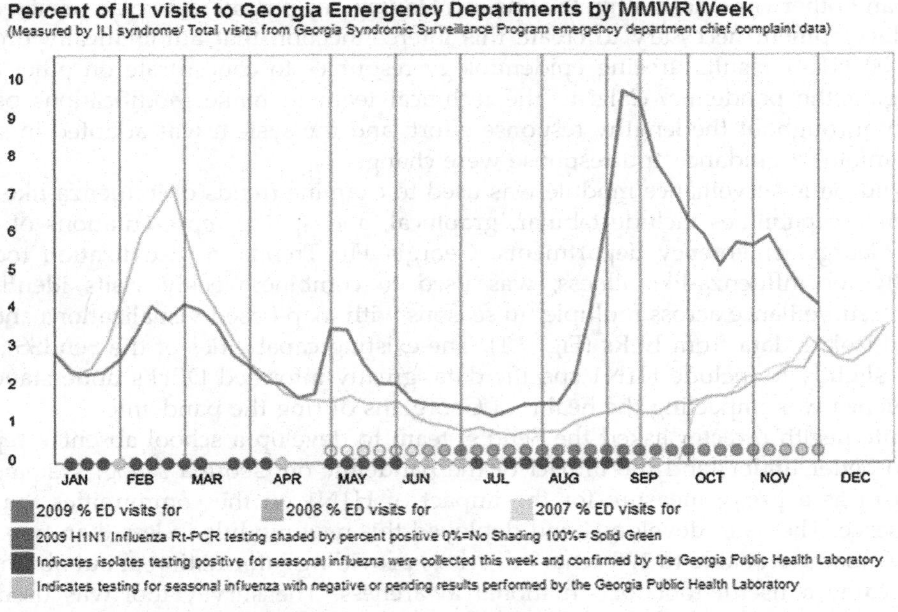

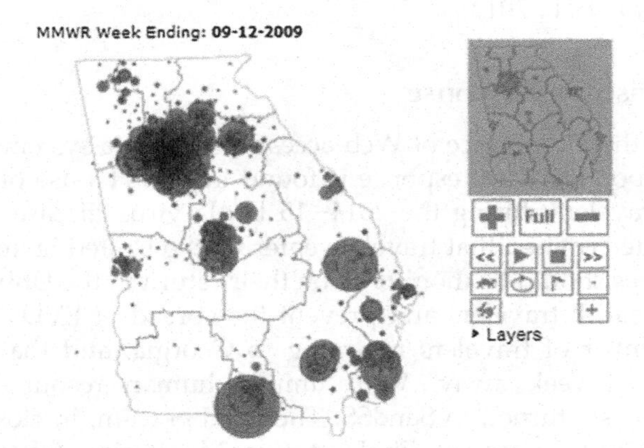

FIGURE 3.1 Georgia Flu Trends dashboard demonstrates the time series of weekly emergency department and urgent care visits for influenza-like illness (ILI) across multiple flu seasons and viral isolate data from 2009. The map demonstrates the weekly geographic distribution of ILI visits.

(Fig. 3.2); and (4) an online query capability designed to enable emergency departments to search EAMS by traveler name and date of birth to quickly determine whether a patient was enrolled in active monitoring. Epidemiologists in Georgia's 18 health districts could log into the system to view and follow up with travelers in their own district, and designated CDC personnel had a customized view to help in the management of their

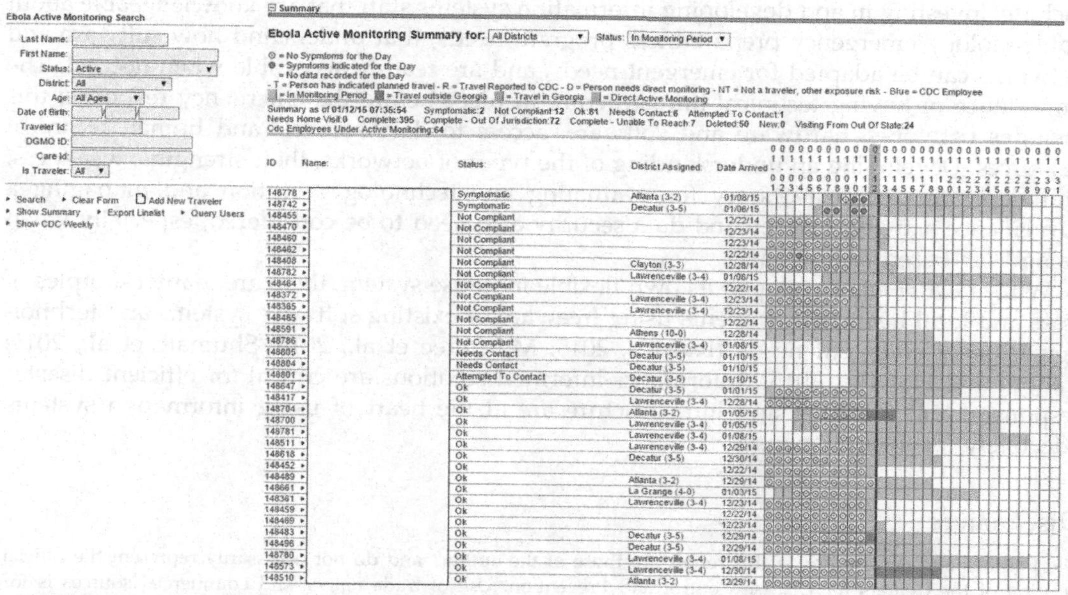

FIGURE 3.2 The Ebola Active Monitoring System dashboard displayed a line list of travelers, summary statistics, information about the travelers, a graph of the active monitoring period, reported symptoms, and a search engine.

Georgia-based employees returning from deployment to EVD-affected countries. The use of EAMS significantly reduced the number of person-hours required to manage this intensive active surveillance activity and allowed two epidemiologists to monitor more than 130 travelers daily, or 2960 total travelers. In addition, EAMS enabled easy documentation and communication across state, local, and federal partners when travelers became ill and needed medical follow-up; 42 ill travelers were medically evaluated and followed closely by public health. The flexibility of the system, as evidenced by its ability to serve travelers for symptom entry, state and district staff for the administration of records, and CDC personnel to monitor their staff, was crucial to the response and made possible by the use of SendSS Web-based technology and the efforts of accessible skilled technical personnel (Parham et al., 2015).

BUILDING INFRASTRUCTURE

The GDPH examples provided in this chapter provide a brief overview of the ways information systems can be used to augment surveillance and other functions of an emergency response.

Building the foundations for flexible surveillance systems is critical for adapting to unpredicted or emergent events. The demonstrated success of SendSS as used in response to these public health threats stems in large part from the strength of its foundation. Items to consider include: investing in and developing information systems staff that are knowledgeable about epidemiology emergency preparedness program needs, that understand how software and networks can be adapted for emergent needs, and are readily accessible when needed. The importance of having technical infrastructure in place before the emergency response; this includes networks, hardware and software, access to data sources, and human technical resources. Developing an understanding of the types of networks, their strengths, weakness, and dependencies is necessary for evaluating any technology solution and managing a surveillance activity. Privacy and data security also need to be considered, especially when using Web-based technology.

While GDPH has developed its own flexible in-house system, there are many examples of state and local health departments using freeware or existing software systems and technology to meet emergent needs (Harney, 2015; Murphree et al., 2014; Shumate et al., 2015; Vahora and Hoferka, 2015). Informatics-informed solutions are critical for efficient disaster responses, and planning and infrastructure are at the heart of using information systems effectively for these responses.

Disclaimers

The findings and conclusions in this report are those of the authors and do not necessarily represent the official position of the Centers for Disease Control and Prevention. Use of trade names and commercial sources is for identification only and does not imply endorsement by the Centers for Disease Control and Prevention, the Public Health Service, or the U.S. Department of Health and Human Services.

References

Atrubin, D., Wiese, M., Snider, R., Workman, K., & Mcdougle, W. (2013). Enhanced disease surveillance during the 2012 republican national convention, Tampa, FL. *Online Journal of Public Health Informatics, 5*(1). http://dx.doi.org/10.5210/ojphi.v5i1.4558.

Ayala, A., Berisha, V., Goodin, K., Pogreba-Brown, K., Levy, C., Mckinney, B., et al. (2016). Public health surveillance strategies for mass gatherings: Super bowl XLIX and related events, Maricopa county, Arizona, 2015. *Health Security, 14*(3), 173–184. http://dx.doi.org/10.1089/hs.2016.0029.

Baker, K. (May 13, 2015). Retrieved from http://c.ymcdn.com/sites/www.cste.org/resource/resmgr/DisasterEpi/Baker_Disaster_Epi_Conferenc.pdf.

Harney, S. (June 15, 2015). Retrieved from https://cste.confex.com/cste/2015/webprogram/Paper4370.html.

Howland, R. E., Li, W., Madsen, A. M., Wong, H., Das, T., Betancourt, F. M., et al. (2015). Evaluating the use of an electronic death registration system for mortality surveillance during and after Hurricane Sandy: New York City, 2012. *Am J Public Health American Journal of Public Health, 105*(11). http://dx.doi.org/10.2105/ajph.2015.302784.

Lombardo, J., Sniegoski, C., Loschen, W., Westercamp, M., Wade, M., Dearth, S., et al. (2008). Public health surveillance for mass Gatherings. *Johns Hopkins APL Technical Digest, 27*(4), 347–355.

Murphree, R., Petersen, P., Stanford, L., Sexton, J. C., Mckennley, T., Milliken, J., et al. (2014). Leveraging public health emergency informatics during the fungal infections outbreak, Tennessee – 2012. *Online Journal of Public Health Informatics, 6*(1). http://dx.doi.org/10.5210/ojphi.v6i1.5125.

Nasrullah, M., Breiding, M. J., Smith, W., Mccullum, I., Soetebier, K., Liang, J. L., et al. (2012). Response to 2009 pandemic influenza A H1N1 among public schools of Georgia, United States—fall 2009. *International Journal of Infectious Diseases, 16*(5). http://dx.doi.org/10.1016/j.ijid.2012.01.010.

Parham, M., Edison, L., Soetebier, K., Feldpausch, A., Kunkes, A., Smith, W., et al. (2015). Ebola active monitoring system for travelers returning from West Africa—Georgia, 2014–2015. *Morb Mortal Wkly Rep, 64*(13), 347–350.

Rodgers, L. E., Paulson, J., Fowler, B., & Duffy, R. (2015). System for rapid assessment of pneumonia and influenza-related mortality—Ohio, 2009–2010. *American Journal of Public Health, 105*(2), 236–239. http://dx.doi.org/10.2105/ajph.2014.302231.

Shumate, A. M., Yard, E. E., Casey-Lockyer, M., Apostolou, A., Chan, M., Tan, C., et al. (2015). Effectiveness of using cellular phones to transmit real-time shelter morbidity surveillance data after Hurricane Sandy, New Jersey, October to November, 2012. *Disaster Medicine and Public Health Preparedness, 10*(03), 525–528. http://dx.doi.org/10.1017/dmp.2015.164.

Vahora, J., Hoferka, S. (2015, June 11). Retrieved from http://www.cste.org/blogpost/1084057/219374/How-Illinois-Used-REDCap-to-Support-Contact-Monitoring-for-the-2015-Measles-Outbreak.

CHAPTER

4

Applications: Shelter Surveillance

Rebecca J. Heick

Augustana College, Rock Island, IL, United States

INTRODUCTION

Surveillance is, by definition, an ongoing system of data collection to monitor the health status of populations and serves as an early warning system for potentially harmful events. In disaster settings, where populations are not only under extraordinary stress but are also placed into an unanticipated communal living situation, surveillance becomes critical to our ability to guard the health of both displaced populations and those serving them. Health surveillance in shelters is critical for both the immediate response to identified needs and for resource allocation for the projected life span of the shelter and beyond. The data collected can also be useful for communicating with partners and the media, as well as in preparedness planning for future responses. A good shelter health surveillance system will provide early identification of potential problems and ultimately result in a safer, healthier living environment for those housed there. While most of the issues identified via surveillance will not require an emergent response, early, rapid identification of these issues does result in improved intervention and increased success in maintaining an optimal health environment in the shelter.

WHY CONDUCT SHELTER SURVEILLANCE?

Shelter surveillance provides information in a timely, organized fashion, detailing the overall health status of the housed population as well as identifying potential needs. Surveillance data serve multiple purposes: They provide a depiction of the health status of populations within individual shelters in a geographic area and also help to create a detailed overall snapshot of the health of sheltered populations across multiple facilities. Both of these elements are necessary when discussing our ability to keep residents safe and healthy while displaced and for determining resource allocation during disasters.

Disaster Epidemiology
http://dx.doi.org/10.1016/B978-0-12-809318-4.00005-8

Value of Surveillance in an Individual Shelter

In any individual shelter, surveillance allows for an ongoing picture of the health status of the residents. Each shelter population will have its own unique mix of demographics and health needs. Involving all shelter staff directly in the collection of surveillance data and ensuring data is shared both frequently and consistently will make them aware of and responsive to any changes in status. Good surveillance will identify changes in population health status early, allowing for rapid intervention, which will limit the spread of communicable diseases as well as potentially reduce ongoing exacerbation of chronic conditions within the population (Noe et al., 2013). However, the need for rapid intervention and optimal resource allocation goes beyond the walls of a single shelter and requires that all shelters be able to consistently aggregate and share health data.

Value of Surveillance Across Multiple Shelters

While individual shelter data are valuable, it must be incorporated into a larger surveillance system that collects data from all shelters involved in the disaster response. The ability to rapidly aggregate and distribute surveillance data from multiple shelters is key in protecting the health of populations. If one shelter is experiencing an increase in syndromic symptoms such as fever or diarrhea, it is important to know if this is an isolated occurrence, or part of a larger health issue occurring across multiple shelters. Increased reports of syndromic symptoms in shelters will be shared with local public health agencies or health systems so that they know the full scope of the issue in these displaced populations. Review of data across all shelters in a response also allows for adjustment of health resource allocation should a specific shelter(s) have a high need population.

Goals of Shelter Surveillance

Health surveillance in shelters is conducted to rapidly identify potential threats to the health of the residents and maintain the healthiest possible environment. These threats often include infectious diseases that are easily spread in communal settings—gastrointestinal illnesses such as norovirus and respiratory infections such as influenza are frequent in shelter settings (Cookson et al., 2008; Murray et al., 2009). Beyond the identification of immediate health threats, shelter surveillance provides a good opportunity to identify longer-term needs for these populations and is useful in the development of prevention messages for current and future disasters. Achieving these goals requires a thoughtful, organized approach to the data collection process.

When to Initiate Health Surveillance in Shelters

One of the major questions in establishing a surveillance system is the decision to activate it. On the surface, it seems simple—if a shelter is open, surveillance should be occurring. As we look deeper into the issue, it becomes clear that shelter surveillance may be an activity we engage in only under specific circumstances. If a single shelter is open and housing eight displaced individuals (anticipated to need housing for 48 h or less), the need for surveillance will be minimal. However, this scenario could be leveraged to proactively validate surveillance systems and processes within the shelter. The decision to initiate surveillance in shelters depends on the circumstances and available resources. As each disaster presents a unique scenario, shelter-specific trigger points for the initiation of health surveillance should be identified and included in the shelter's surveillance plan.

DATA COLLECTION

To collect data to rapidly identify immediate threats, guide resource allocation, and have utility in generating prevention messages, methods must be developed that can be consistently carried out in shelter settings with only minor modifications across locations and events. Methods for surveillance are placed into two primary categories: active surveillance (e.g., health information is sought out from shelter residents in cot-to-cot rounds) or passive surveillance (e.g., health information is gathered when a shelter resident approaches a health services staff member with a concern). Each approach requires a specific set of skills and resources to be successful, making one approach more practical than the other in many circumstances. A combination of the two approaches is often most successful.

Active Surveillance

Active surveillance involves cot-to-cot rounds as a method of touching base with all residents in a shelter to inquire about how they are feeling physically and mentally and to determine if they have any health concerns or unmet needs. This approach is especially effective with large shelter populations as it sends staff members throughout the entire shelter to seek residents out. When conducting cot-to-cot surveillance, it is important to divide up the shelter into sectors to ensure the maximum numbers of residents are contacted with the least duplication. Data gathered from this type of surveillance are less detailed than those of individual health records (e.g., often including only age group and reported symptoms), but it is a valid method to collect surveillance data and is widely used (Murray et al., 2009). These cot visits are ideally completed by health services staff, but can also be completed by trained volunteers or nonmedical shelter staff with an understanding of when to engage health services. Cot-to-cot rounds may result in the generation of an individual health record when a resident is referred to health services staff or when these staff are requested to come to the resident. Therefore, surveillance process should account for the potential duplication in reporting of symptoms. One method for doing this is to have an area on the cot-to-cot form which tracks referrals to health services and the generation of an individual health record, as well as the chief complaint or symptom that generated the referral.

Passive Surveillance

Passive surveillance is typically based on individual contacts with health services personnel and is most often initiated by the shelter resident who seeks out health services staff because they do not feel well or have a concern. The individual health record from passive surveillance generally includes a significant amount of data, including demographics, chief complaint, and past medical history. Passive surveillance will often generate a more limited number of records than active surveillance, but the reduced quantity of data is often balanced with an increase in the depth and detail of what is recorded. This demonstrates the potential need for both types of surveillance in shelter settings.

Data Collection Format Considerations

Before forms are designed for shelter health surveillance processes, it is critical to consider the appropriate format and back-up mechanism. The circumstances under which these forms are used are rarely optimal—there may be issues with electricity, connectivity, or reproduction. Paper forms are an excellent option when circumstances with no power or connectivity are considered—provided that sufficient stores (enough for at least 3 to 4 days initially) are available in each facility in case copying is not readily available. However, paper forms also limit the shelter's ability to transmit the data directly to the facility where aggregation will occur. Good options for submitting data from paper forms include faxing, phoning in the numbers or sending it as a picture via text or email. In addition, one must consider that forms may need to be updated regularly, which will require all stored forms to be swapped for new versions. Continued use of outdated versions may create additional challenges for data aggregation and reporting. Paper forms typically require handwritten responses, which may be difficult to decode. Taking into account these limitations, electronic versions should also be considered.

Electronic forms have several advantages over paper, including the ease of sending and updating. Entries are more consistent (and more legible) because limits can be placed on the fields, catching potentially erroneous entries prior to submission. In some instances, real-time reporting can occur with only minimal lag between submission from individual shelters and the availability of aggregate data. The major drawback to electronic reporting is the challenge of connectivity and the necessity of a good power supply. Neither of these resources is guaranteed during a disaster, and the time required to reestablish these services can vary significantly from one disaster response to the next. Publicly available electronic platforms such as CDC's Epi Info, a free product that works on computer and mobile devices and via the web or cloud, warrant consideration.

The best recommendation when making a decision on data collection format is to consider the technical literacy of those who are likely to handle the data collection and reporting. What will they feel most comfortable with? What is the likelihood that smart phones, tablets, or computers will be available at all shelter locations? Although electronic formats are attractive options, digital records should be duplicated on paper to maintain accessibility in the event of power loss or network outages. Failure to plan for the inevitable disruptions of a disaster can result in a dysfunctional surveillance system when it is critically needed.

DATA AGGREGATION

Aggregating Data at the Shelter Level

When data are collected within a single shelter, either through passive or active surveillance methods, it is important to have a process in place that allows for data to be aggregated on a regular basis. The aggregation of data requires consistency across all the documents used to ensure they are fully compatible (e.g., categories are the same on all the documents). The timeliness and frequency of reporting to decision-makers or leadership outside the shelter should also be considered when developing data aggregation protocols. Setting an established timeframe for data aggregation ensures that there is a consistent time span for each reporting period. A midafternoon report time often works well since daytime staffing is usually adequate to let someone step away to do the work, and this avoids the chaos of mealtimes. The reporting time should be the same for all shelters (Fig. 4.1).

Aggregating Data Across Shelters

Once data are collected at the individual shelter level, a process must exist for aggregation of the data into a single document or report that can then be shared back to the shelters and with local health officials and public health agencies. If forms present in the shelters are consistent in design, the actual process of data aggregation is fairly straightforward. It simply becomes a matter of going through each form and inputting the data to generate category totals for the reporting period. The challenge of aggregating data from multiple disaster shelters lies in the process of getting all the forms to a single location in a timely fashion so that aggregation can occur.

Reporting Aggregate Data

The first step in the process is determining what format the information will be in. As discussed previously, there are advantages and disadvantages to both paper and electronic platforms. Knowing the disaster response team and its capabilities will make this decision much easier. Once the required forms are created and a platform has been selected, a timeframe for reporting will need to be established. Individual shelters should also know when to expect information back each day (Fig. 4.2). The format of the aggregate report is also important so that those receiving the daily report know what to expect and where to look in the report for various items. Graphics can be extremely helpful in comparing current data to prior day(s), allowing an at-a-glance picture of the health status of the displaced population. Determining how the report will be distributed (e.g., email, fax, hard copy) in advance will also impact the overall format and use of graphics.

A final major consideration for shelter health surveillance data is with whom it should be shared. It is given that each shelter who reported data should receive an aggregate report, but what about local health and public health agencies? The best way to address this question is to reach out to these agencies very early on in the sheltering operation (or even prior to

American Red Cross

Disaster Health Services Aggregate Morbidity Report Form*

Part I. General Information
1. Disaster Operation # ________________
2. Reporting Date: ___/___/___
3. Reporting Timeframe: ______ – ______
4. County ________________ State ____
5. Service Type (circle): Shelter Non-Shelter
6. Worksite Name: ________________

Part II. Number of Client-Related Interactions Tally (卌 卌 卌)	Total (#)
7. Total Client-related Contacts:	
7b. Total of Health-related Client Visits: (fill part III)	

Part III. Demographics (for Health-related Visits Only)		Tally (卌 卌 卌)	Total (#)
Gender	Male		
	Female		
Age	≤ 2		
	3 to 18		
	19 to 64		
	≥ 65		

Functional/Access Needs: mark each individual need based on C-MIST model per 24 hours	Tally (卌 卌 卌)	Total (#)
Communication		
Maintenance of Health		
Independence		
Safety and Security		
Transportation		

Part IV. Reason for Visit: for each client visit, tick ALL reason(s) for visits.

	Tally (卌 卌 卌)	Total (#)
Injury		
Bite (includes *ALL* bites)		
Burn (thermal or chemical)		
Cut/laceration/puncture		
Foreign body (e.g., splinter)		
Fall/slip/trip		
Hit by or against object		
Use of machinery/tools/equip.		
Assault		
Carbon Monoxide (CO) exposure		
Poisoning, non-CO		
Other injury		
Illness/Symptoms		
Fever (>100.4°F or 38°C)		
Conjunctivitis/eye irritation		
Dehydration		
Heat stress/heat exhaustion		
Hypothermia/cold-environment		
Oral health		
Pain: chest, angina, cardiac arrest		
Pain: muscle or joint pain		
Pain: head, ears, eyes, nose, throat		
Pain: other, not specified above		
Gastrointestinal (GI): diarrhea		
GI: nausea/vomiting		
GI: other (constipation, GERD)		
Genitourinary (GU)		
Skin (includes *ALL* skin conditions)		
Allergic reaction		
Respiratory (include *ALL* resp.)		
Influenza-like-illness (ILI)		
Neurological, new onset		
Other illness/symptoms		

	Tally (卌 卌 卌)	Total (#)
Behavioral/Mental Health		
Agitated/disruptive/psychotic		
Anxiety/stress/depressed mood		
Suicidal/homicidal thoughts		
Substance addiction/withdrawal		
Other mental health		
Exacerbation of Chronic Illness		
Asthma		
Obstructive pulmonary disease		
Cardiovascular (HTN, CHF, CHD)		
Chronic muscle or joint pain		
Diabetes		
Neurological (seizure, stroke, dementia)		
Previous mental health diagnosis		
Other chronic illness		
Health Care Maintenance		
Blood pressure check		
Blood sugar check		
Pregnancy/post-partum care		
Dressing change/wound care		
Immunization/vaccination		
Medical refill (please mark one tick for *each* med refill)		
Other health maintenance		

Part V. Disposition	Tally (卌 卌 卌)	Total (#)
Provided Red Cross care		
Referred to...		
Hospital		
Physician/dentist/clinic		
Pharmacist		
Other (e.g., DMH)		
Refused Red Cross care		

***Complete one form per service location per 24 hours. Submit by 4pm local time.**

Print name: ________________ Contact information: ________________ Aggregate Morbidity Report Form 2077C (rev. 2/13)

FIGURE 4.1 Disaster health services aggregate morbidity report form.

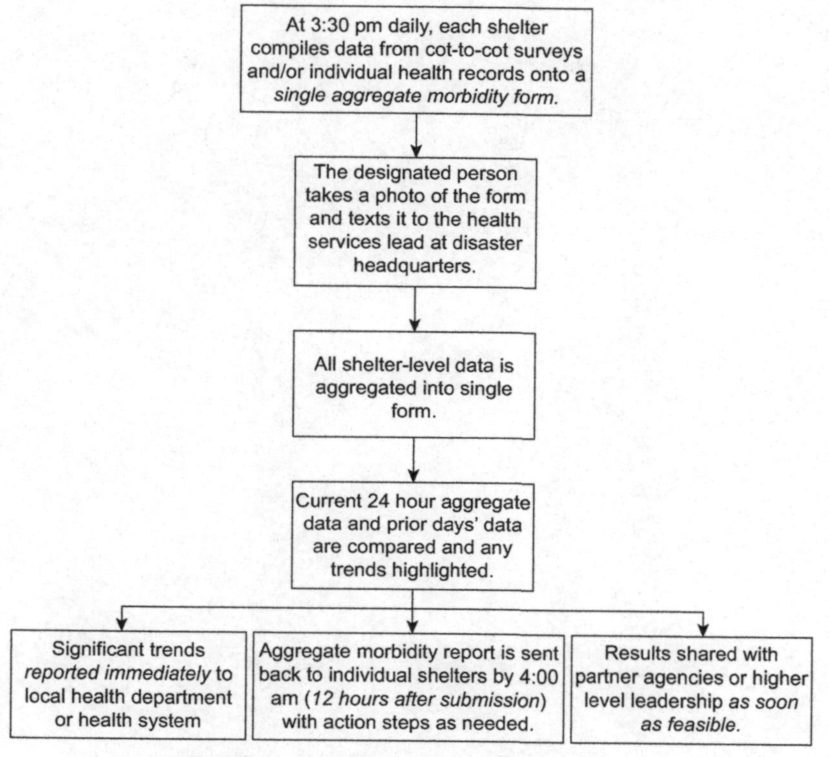
Sample Shelter Surveillance Process
24 Hour Reporting Period: 4:00 pm to 4:00 pm
At 3:30 pm daily, each shelter compiles data from cot-to-cot surveys and/or individual health records onto a single aggregate morbidity form.
The designated person takes a photo of the form and texts it to the health services lead at disaster headquarters.
All shelter-level data is aggregated into single form.
Current 24 hour aggregate data and prior days' data are compared and any trends highlighted.
Significant trends reported immediately to local health department or health system
Aggregate morbidity report is sent back to individual shelters by 4:00 am (12 hours after submission) with action steps as needed.
Results shared with partner agencies or higher level leadership as soon as feasible.

FIGURE 4.2 Sample shelter surveillance process.

the event if possible) to find out what information they would like to receive, how frequently, and to whom it should go to within the agency. Assigning one person to handle the dissemination of the daily aggregate information will streamline the process and help maintain relationships with outside agencies.

References

Cookson, S. T., Soetebier, K., Murray, E. L., Fajardo, G. C., Hanzlick, R., Cowell, A., et al. (2008). Internet-based morbidity and mortality surveillance among Hurricane Katrina evacuees in Georgia. *Preventing Chronic Disease, 5*(4).

Murray, K. O., Kilborn, C., DesVignes-Kendrick, M., Koers, E., Page, V., Selwyn, B. J., et al. (2009). Emerging disease syndromic surveillance for Hurricane Katrina evacuees seeking shelter in Houston's Astrodome and Reliant Park Complex. *Public Health Reports, 124*, 364–371.

Noe, R. S., Schnall, A. H., Wolkin, A. F., Podgornik, M. N., Wood, A. D., Spears, J., et al. (2013). Disaster related injuries and illnesses treated by American red cross disaster health services during Hurricanes Gustav and Ike. *Southern Medical Journal, 106*(1), 102–108.

Vignette: Postdisaster Carbon Monoxide Surveillance

Kanta Sircar, Dorothy Stearns

Centers for Disease Control and Prevention, Chamblee, GA, Unites States

CARBON MONOXIDE POISONING IN THE AFTERMATH OF HURRICANE SANDY

Hurricane Sandy made landfall on the New Jersey coast on October 29, 2012. New Jersey and several other Northeastern US states were impacted by storm surges and flooding that led to widespread power outages, damaged infrastructure, and deaths. The storm left approximately 8.5 million homes and businesses without power (Bryan, 2012). To cope with storm-related power losses and extremely cold temperatures, people sometimes resorted to using generators and carbon fuel–based products to produce electricity and heat inside their homes. The use of these devices without proper ventilation allowed carbon monoxide (CO) levels to increase, exposing occupants to the poisonous gas. As in other postdisaster situations, risk-taking behavior among survivors increases when power or heat is a limited resource. After the storm, regional poison centers in eight states affected by Hurricane Sandy reported an increase in unintentional CO poisonings (Centers for Disease Control and Prevention, 2012). Emergency departments (EDs) also reported increased numbers of CO poisonings during the storm period.

WHAT IS CARBON MONOXIDE POISONING?

CO poisoning is a preventable cause of death and a leading cause of morbidity in postdisaster situations. With the exception of alcohol, CO is responsible for more deaths each year than any other nonmedical toxicant (Sircar et al., 2015). CO is an odorless and colorless gas that is produced as a result of the incomplete combustion of carbon-based fuels. When inhaled, CO gas binds to hemoglobin in red blood cells, reducing the blood's ability to carry oxygen throughout the body. In addition, this exposure impairs cellular utilization and respiration of oxygen. CO poisoning is dose dependent; symptoms worsen as the concentration of

Disaster Epidemiology
http://dx.doi.org/10.1016/B978-0-12-809318-4.00006-X

TABLE 1 Symptoms Reported to the National Poison Data System by Carbon Monoxide-exposed* Individuals, 2000–09 (n = 68,316)

	Number	Percentage (%)
Headache	30,845	45.2
Nausea	17,653	25.8
Dizziness/vertigo	13,363	19.6
Drowsiness/lethargy	8966	13.1
Vomiting	7550	11.1
Confusion	2083	3.0
Syncope	1950	2.9
Dyspnea	1538	2.3
Chest pain	1226	1.8
Other	6548	9.6
None	21,793	31.9

** Includes unintentional, nonfire-related exposure.*

Adapted from Bronstein, A., Clower, J. H., Iqbal, S., Yip, F. Y., Martin, C. A., Chang, A., et al. (2011). Carbon monoxide exposures – United States, 2000–2009. Morbidity and Mortality Weekly Report, *60(30), 1014–1017.*

CO in the blood rises. Nonspecific clinical effects can include headache, nausea, vomiting, dizziness, and loss of consciousness (Table 1); however, CO poisoning may occur in asymptomatic individuals. Long-term effects can include delayed neurologic sequelae (which can occur more than a month after initial CO exposure), coma, cardiovascular disease, and death (Centers for Disease Control and Prevention, 2008). One treatment method for severe cases is hyperbaric oxygen therapy, which effectively supplies oxygen to the body.

WHY IS CARBON MONOXIDE POISONING A PUBLIC HEALTH CONCERN?

The National Center for Environmental Health at the Centers for Disease Control and Prevention (CDC) conducts research and surveillance in collaboration with local and state health departments on the detection and prevention of CO poisoning cases. Each year, approximately 438 deaths and more than 21,000 ED visits occur as a result of unintentional CO exposure in the United States (Centers for Disease Control and Prevention, 2008; Iqbal, Clower, Hernandez, Damon, & Yip, 2012; Sircar et al., 2015). Many unintentional, nonfire-related CO exposures result from improper placement of carbon-fueled products, such as generators and charcoal grills, inside the home. For example, a standard, gasoline-powered 5.5 kW portable generator for home use produces about the same amount of CO as six idling cars (Centers for Disease Control and Prevention, 2005), which may lead to poisonings when these devices are placed near or inside the home to avoid theft or wet conditions. In a CDC study, generator use accounted for 83% of CO fatalities and 54% of nonfatal exposures (Iqbal, Clower, Hernandez, Damon, & Yip, 2012). Through the use of legislative mandates, some local and state authorities now require the placement of battery-operated CO detectors in homes (Graber, Macdonald, Kass, Smith, & Anderson, 2007).

ESTABLISHING CASES OF CARBON MONOXIDE EXPOSURE

Data for retrospective surveillance of CO poisoning are obtained from case reports, medical databases, and death certificates. Exposures of interest include acute or suspected contacts with CO, but poisoning cases are stronger exposures that result in an adverse health response. Prospective surveillance for likely CO poisoning cases can also be done using information from various sources. These include records of ED and hospital visits, use of hyperbaric oxygen therapy, and calls to poison centers; laboratory, medical examiner, and coroner reports; medical databases; and questionnaires. Note that some hospitals might close during a disaster situation, which can affect the timeliness and accuracy of reporting.

National routine case notification through the National Poison Data System of the American Association of Poison Control Centers has improved the surveillance of CO exposures (Wolkin, Martin, Law, Schier, & Bronstein, 2012). Case definitions for CO poisoning have been established by the Council of State and Territorial Epidemiologists (CSTE) and can be accessed through the CSTE webpage. This system uses suggested criteria for identifying cases, such as clinical presentation, laboratory, and case finding in administrative data criteria. However, these systems might not capture all CO poisoning cases. Persons with mild symptoms of CO exposures might not seek medical attention or might be misdiagnosed with other illnesses, resulting in underreporting of CO poisoning cases.

THE EPIDEMIOLOGY OF CARBON MONOXIDE POISONING AFTER A DISASTER

CO poisonings most often occur after ice storms, hurricanes, floods, and power outages. CO exposure can result from various sources; these include charcoal grills, portable gas stoves, space heaters, kerosene heaters, and gasoline-powered electric generators inside the home during or after a storm. No two disasters are alike and CO poisoning risks vary depending on the type of disaster. For example, in winter storms, snow drifts can block home heating vents or car exhaust pipes, causing a buildup of CO inside homes or vehicles (Johnson-Arbor, Quental, & Li, 2014). Older catalytic converters pose a risk during a snowstorm for potential CO accumulation. The CO poisoning risk from power outages is similar to that of other disaster types because most disasters disrupt electrical service. Other disasters, such as floods and hurricanes, can lead to accidental poisonings as a result of using gasoline-powered equipment to clean up postdisaster or using flood-damaged equipment months after the disaster (Waite, Murray, & Baker, 2014). These exposures can occur days, weeks, or months after a storm or power outage, making it difficult to quantify the extent of CO poisoning resulting from the event.

Children generally have a greater risk for CO exposures that result in an ED visit. Although the greater proportion of persons exposed to CO are women, men make up the majority of deaths from CO poisoning. This difference might reflect a greater proportion of men engaging in high-risk activities involving the use of generators and heaters (Centers for Disease Control and Prevention, 2009). Women typically show exposure symptoms at lower CO levels because they have lower red blood cell count, leading to earlier recognition (Iqbal, Clower, Boehmer, Yip, & Garbe, 2010).

Excess morbidity from CO exposure among non-English speaking minorities has been reported in multiple studies. Indoor use of charcoal grills or solid fuel for cooking or heating might be associated with people of African, Middle Eastern, or Asian origin (Daley, Smith, Paz-Argandona, Malilay, & McGeehin, 2000). For example, out of 44 persons poisoned from charcoal briquette use after a 1993 winter storm in Washington State, 40 were members of ethnic minorities (Houck & Hampson, 1997). Among all patients, 66% of the members of these ethnic minority groups did not speak English. In another example, Hispanics accounted for 65% of CO exposures in a 2002 North Carolina ice storm, but only 5% of the state's population (Broder, Mehrotra, & Tintinalli, 2005). Instructions for fuel-powered devices, such as generators or charcoal grills, might be printed in small type or might only be available in English.

HOW TO PREPARE?

Predisaster risk communication should reflect the diversity of its intended audience, and intervention strategies should be tailored to different types of storms. One way to reduce CO poisoning is good predisaster communication. As part of that, media outlets might provide CO exposure prevention information. CDC's National Disasters and Severe Weather webpage provides prevention tips and brief videos on CO poisoning prevention as well. Easy to understand language and diagrams on packaging and user manuals might help reduce exposures from products that generate CO. Increased education for health-care providers on recognizing CO symptoms and legislation promoting CO monitor use in homes are additional ways to address the problem. All these issues highlight the need for an active and comprehensive CO surveillance system, which provides environmental hazard information, increased notification and tracking of CO exposures, and effective intervention strategies to reduce CO poisonings.

References

Broder, J., Mehrotra, A., & Tintinalli, J. (2005). Injuries from the 2002 North Carolina ice storm, and strategies for prevention. *Injury, 36*(1), 21–26. http://dx.doi.org/10.1016/j.injury.2004.08.007.

Bronstein, A., Clower, J. H., Iqbal, S., Yip, F. Y., Martin, C. A., Chang, A., et al. (2011). Carbon monoxide exposures – United States, 2000–2009. *Morbidity and Mortality Weekly Report, 60*(30), 1014–1017.

Bryan, W. N. (2012). *Hurricane Sandy situation report #20. Washington.* Retrieved from http://www.oe.netl.doe.gov/docs/2012_SitRep20_Sandy_11072012_1000AM.pdf.

Centers for Disease Control and Prevention. (2005). Carbon monoxide poisoning from hurricane-associated use of portable generators—Florida, 2004. *Morbidity and Mortal Weekly Report, 54*, 697–700.

Centers for Disease Control and Prevention. (2008). Nonfatal, unintentional, non-fire-related carbon monoxide exposures—United States, 2004–2006. *Morbidity and Mortal Weekly Report, 57*, 896–899.

Centers for Disease Control and Prevention. (2009). Carbon monoxide exposures after Hurricane Ike—Texas, September 2008. *Morbidity and Mortal Weekly Report, 58*, 845–849.

Centers for Disease Control and Prevention. (2012). Notes from the field: Carbon monoxide exposures reported to poison centers and related to Hurricane Sandy—northeastern United States, 2012. *Morbidity and Mortal Weekly Report, 61*, 905.

Daley, W. R., Smith, A., Paz-Argandona, E., Malilay, J., & McGeehin, M. (2000). An outbreak of carbon monoxide poisoning after a major ice storm in Maine. *Journal of Emergency Medicine, 18*, 87–93.

Graber, J. M., Macdonald, S. C., Kass, D. E., Smith, A. E., & Anderson, H. A. (2007). Carbon monoxide: The case for environmental public health surveillance. *Public health reports*, 138–144.

Houck, P. M., & Hampson, N. B. (1997). Epidemic carbon monoxide poisoning following a winter storm. *Journal of Emergency Medicine, 15*, 469–473.

Iqbal, S., Clower, J. H., Boehmer, T. K., Yip, F. Y., & Garbe, P. (2010). Carbon monoxide-related hospitalizations in the US: Evaluation of a web-based query system for public health surveillance. *Public Health Reports*, 423–432.

Iqbal, S., Clower, J. H., King, M., Bell, J., & Yip, F. Y. (September–October 2012). National carbon monoxide poisoning surveillance framework and recent estimates. *Public Health Reports, 127*

Johnson-Arbor, K. K., Quental, A. S., & Li, D. (May 2014). A comparison of carbon monoxide exposures after snowstorms and power outages. *American Journal of Preventive Medicine, 46*(5), 481–486. http://dx.doi.org/10.1016/j.amepre.2014.01.006.

Sircar, K., Clower, J., Kyong Shin, M., Bailey, C., King, M., & Yip, F. (2015). Carbon monoxide poisoning deaths in the United States, 1999 to 2012. *The American Journal of Emergency Medicine, 33*(9), 1140–1145.

Waite, T., Murray, V., & Baker, D. (2014). Carbon monoxide poisoning and flooding: Changes in risk before, during and after flooding require appropriate public health interventions. *PLoS Currents, 6*. http://dx.doi.org/10.1371/currents.dis.2b2eb9e15f9b982784938803584487f1.

Wolkin, A. F., Martin, C. A., Law, R. K., Schier, J. G., & Bronstein, A. C. (2012). Using poison center data for national public health surveillance for chemical and poison exposure and associated illness. *Annals of Emergency Medicine, 59*, 56–61. http://dx.doi.org/10.1016/j.annemergmed.2011.08.004.

CHAPTER

5

Applications: Disaster-Related Mortality Surveillance

Challenges and Considerations for Local and State Health Departments

Rebecca S. Noe

Centers for Disease Control and Prevention, Atlanta, GA, United States

BACKGROUND

Approximately 2000 people in the United States die each year from severe weather-related events (Berko, Ingram, Saha, & Parker, 2014). Of these, the majority of deaths are attributable to major floods, tornadoes, and exposure to excessive environmental heat and cold (Berko et al., 2014). Accurate enumeration of decedents after a disaster can provide insight into the severity of the event and its burden on the affected population. Data about decedents' demographics, locations of injury, and circumstances of deaths can be used to identify the leading cause(s) of death, at-risk groups, and common risk factors of the disaster event. Information on deaths guide the public health response, helping to determine the appropriate type and amount of resources to request and mobilize, such as deployment of a Federal Disaster Mortuary Operational Team (DMORT) or strategies to address ongoing hazards (e.g., public health messaging). In addition, accurate mortality data can inform and support preparedness policies and assist with planning for future disasters.

Any approach to conducting disaster-related mortality surveillance requires considerable planning and coordination of the numerous partners that are involved in public health emergency response and interested in mortality information. To assist with planning the approach, the Centers for Disease Control and Prevention (CDC) created *A Primer for Understanding the Principles and Practices of Disaster Surveillance in the United States* that provides introductory information on the purpose, importance, and framework for approaching disaster surveillance (CDC, 2016). Disaster-related mortality surveillance efforts use a variety of data sources and existing or ad hoc systems. Data sources can

http://dx.doi.org/10.1016/B978-0-12-809318-4.00007-1

include medical examiner (ME) reports, hospital records, and death certificates, with the type of data used dependent on the disaster phase (Table 5.1). For example, mortality data are used during recovery to monitor affected communities for hazards associated with cleanup after the disaster has passed. During the mitigation and preparedness phases, researchers use mortality data to investigate risk factors for death or evaluate interventions to determine their effectiveness at protecting people during a disaster. Findings can help develop evidence-based prevention strategies to mitigate mortality in future disasters.

There are several fundamental challenges and important considerations that impact the design and function of effective mortality surveillance activities during a disaster. This chapter highlights these challenges and presents evidence-based solutions using three case studies.

TABLE 5.1 Data Sources for Disaster-Related Mortality Surveillance

Data Source	Name of Records or/and System	Data Captured
Medicolegal death investigators	*Death Scene Investigation Report*	Details of disaster death scene, circumstances of each of the deaths and risk factors www.abmdi.org and https://www.cdc.gov/nceh/hsb/disaster/docs/DeathSceneInvestigation508.pdf
Medical examiners (MEs), Coroners, justices of peace (JPs)	*Line list* *Case reports*	Data on cause, manner, circumstance of death, determination if death is disaster-related
MEs, Coroners, JPs	*Death Certificate*	Official cause of death, circumstance including disaster-relatedness
Funeral homes	Scene transport records, start/assist in completing death certificate	Transfer of decedent(s), demographics
Emergency medical services	Scene transport records	Transfer of injured to hospital
Hospitals including hyperbaric centers	*Medical record*	Hospital course, hyperbaric treatments, final diagnosis
Vital statistics	Death certificate in Electronic Death Registration System	Official cause of death, circumstance including disaster-relatedness
American Red Cross	*Disaster-related Mortality Report Form*	Demographics, location of event, cause of death, circumstance, and response work-relatedness
News media	Online news reports and memorial sites	Demographics, circumstance, location of event, and risk factors
National Oceanic and Atmospheric Administration (NOAA)-National Weather Service (NWS)	*Storm database*	Overall number of deaths per event, demographics, location of event https://www.ncdc.noaa.gov/stormevents/
Federal Emergency Management Agency	*Individual funeral benefits*	Individual record of benefits distributed https://fema.gov/disater-funeral-assistance

CHALLENGES AND CONSIDERATIONS

The major challenges to conducting disaster-related mortality surveillance can be summarized into three overarching themes: defining disaster-related deaths, reporting disaster-related deaths to public health, and capturing death information in a timely matter.

Challenge: Defining Disaster-Related Deaths

In the late 1990s CDC engaged a group of forensic pathologists, MEs/Coroners, and epidemiologists to develop a matrix for attributing a death to a disaster to improve identification of disaster-related deaths (Combs, Quenemoen, Parrish, & Davis, 1999). The hallmark of this matrix was the definitions for deaths directly and indirectly related to a disaster (Boxes 5.1, 5.2), which provided, for the first time, uniform definitions for the forensic science and public health communities.

These new definitions were applied to deaths during large-scale disasters including Hurricane Katrina and the six hurricanes that made landfall in Florida during the 2004 and 2005 hurricane seasons (Brunkard, Namulanda, & Ratard, 2008; CDC, 2006; Ragan, Schulte, Nelson, & Jones, 2008). However, after Hurricanes Ike and Sandy and the 2011 tornadoes in the southeastern states, a review of death certificates found that the number of deaths attributed to these disasters was grossly under reported (Choudhary et al., 2012; CDC, 2012; CDC personal communications 2011, 2014). The under reporting occurred

BOX 5.1

DIRECTLY-RELATED DISASTER DEATHS (COMBS ET AL., 1999)

Definition: A death directly attributable to the forces of the disaster or by the direct consequences of these forces, such as structural collapse, flying debris, or radiation or chemical exposure.

Questions to ask: *Was the death caused by the actual environmental forces of the disaster such as wind, rain, floods, earthquakes, or blast wave or by the direct consequences of these forces, such as structural collapse, chemical spill or flying debris?*

Common Causes of Direct Disaster-Related Deaths:

Burns
Crushing
Drowning
Electrocution
Falls
Fire/smoke inhalation
Hyperthermia (heat)
Hypothermia (cold)
Radiation/chemical poisoning
Suffocation
Traumatic injury
Blunt force trauma
Penetrating injury

BOX 5.2

INDIRECTLY-RELATED DISASTER DEATHS (COMBS ET AL., 1999)

Definition: A death that occurred due to unsafe or unhealthy conditions present during any phase of the disaster (preevent or preparing for the disaster, during the disaster event, or postevent during cleanup after a disaster).

Questions to ask: *Did unsafe or unhealthy conditions from the environmental forces of the disaster contribute to the death? Did the forces, whether natural or human-induced disaster, lead to temporary or permanent displacement, property damage, or other personal loss or stress that contributed to the death?*

Common Circumstances Leading to Indirect Disaster-Related Death (Combs et al., 1999)

Acute exacerbation of chronic condition(s) (e.g., asthma, cardiovascular)
Cleanup after disaster (e.g., chain saw injury, electrocution)
Escaping or fleeing the disaster (e.g., saw the tornado and fell while rushing down storm shelter stairs, Note: could also be direct if the tornado's winds led to the fall)
Evacuation (e.g., motor vehicle crash while evacuating before the storm)
Exposure to industrial or chemical hazards (e.g., chemical release from hurricane damaged refiners, Note: could also be direct if the exposure was due to a human-induced disaster)
Loss/disruption of public utilities (e.g., fall in home without power)
Loss/disruption of transportation-related services (e.g., lack of medical transport to dialysis)
Loss/disruption of usual access to medical, mental health care (e.g., oxygen)
Preparation for disaster (e.g., fall while putting up storm windows)
Psychosocial stress, anxiety
Social disruption or anarchy
Return to unsafe, unhealthy structures or environment (e.g., electrocution)
Use of temporary sheltering, provisions; displacement

because many of the death certificates did not include key terms, such as "Ike," "Hurricane," "storm surge," "Sandy," "tornado," or "April 27 storm" that identified the death as related to the specific event. It is unclear whether the death was known by the certifier to be related to the specific event and the term was inadvertently omitted from the certificate, or if there was insufficient evidence for the certifier to attribute the death to the specific event. This is an important distinction that highlights the need for death scene investigators, death certifiers (e.g., MEs, Coroners, and justice of the peace (JP)), and others engaged in fatality management during a disaster response to adopt the existing uniform definitions. Increased awareness, understanding, and application of these definitions may improve the type of information collected at the disaster death scene and the accuracy of the certifier to attribute the death to the disaster (Table 5.1).

Indirect deaths are the most difficult to attribute to a disaster. Indirect deaths are not caused by the actual forces of the disaster but are a result of unsafe or unhealthy conditions present during any phase of the disaster (Box 5.2). For example, chronic conditions, such as cardiovascular disease, may be exacerbated by a disaster event and acute medical emergencies, such as electrocution, can occur during postdisaster cleanup (Ragan et al., 2008).

Disaster surveillance plans should include a metric that determines whether a disaster occurred, establishes the start and end date of the disaster, and determines whether surveillance should begin. Information from National Weather Service (NWS) alerts (e.g., warnings, watches, advisories), emergency management orders (e.g., mandatory coastal evacuations, shelter-in-place order), or emergency declarations can help complete the definition metric. The metric also affords the opportunity to establish a unique event term, like "Hurricane Sandy" or "El Derecho—Heat Wave 2012," for death investigators and certifiers to use on the death certificate when attributing and reporting a death caused by the disaster.

To address the challenge in defining a disaster-related death, certifiers should be integrated into the planning for disaster mortality surveillance and together adopt the existing definitions and develop guidelines. These can be shared with all parties involved in fatality management during a disaster and used for training to ensure the accurate and consistent identification of disaster-related deaths.

Challenge: Reporting Information From Deaths Identified as Disaster-Related

Death-Related Information on Death Certificates

Death certificates are the most important source for complete information about disaster-related deaths, but often lack a reference to a specific disaster (e.g., "Hurricane Katrina," "Southwest's Extreme Heat Wave 2016"). This information is important to promptly identify deaths for the immediate public health response and to allow the death to be easily found later when searching databases for deaths related to an event. CDC's *A Reference Guide for Certification of Deaths in the Event of a Severe Weather, Human Induced or Chemical/Radiological Disaster* (https://www.cdc.gov/nchs/nvss/reporting-guidance.htm.) provides certifiers with guidance on how to apply the disaster-related death definitions (Boxes 5.1, 5.2) and properly record disaster-specific information on the death certificate to ensure accurate reporting of these cases.

Additionally, the information written on the death certificate helps in assigning appropriate disaster-related *International Classification of Diseases, 10th Revision* (ICD-10) codes. The ICD-10 codes are ultimately assigned by CDC's National Center for Health Statistics (NCHS). If disaster-related terms (e.g., tornado, flood, Hurricane Sandy) are not recorded on the death certificate, then NCHS will not assign an ICD-10 code that reflects the death as related to a disaster. Therefore, statistics at the state and national level only reflect what was reported on the death certificate. For example, CDC's WONDER public mortality database can be queried for deaths associated with a hurricane (X37.0) captured under the broader code of cataclysmic storm (X37) (http://wonder.cdc.gov/mcd-icd10.html). A query of X37 deaths in New Jersey during the year of Hurricane Sandy (2012), found less than nine X37 deaths despite official Sandy reports of more than 34 disaster-related deaths (CDC, 2013). This demonstrates the importance of including the name and type of the disaster on the death certificate.

Disaster-Related Information Reported Between Key Partners Within the State

Accurate reporting of disaster-related deaths may be impacted by administrative barriers that limit the ability of key partners to share mortality information with public health. Key partners in each state include the certifiers (i.e., MEs, Coroners) and vital statistics offices responsible for collecting and storing mortality data. During a disaster, ME's offices often are responsible for reporting the official number and cause of deaths to response officials and for completing and submitting the death certificates of disaster victims to vital statistics. Once vital statistics receive the death certificates, they may review and analyze them for disaster-relatedness, especially if the disaster is protracted or has substantial media attention leading to requests for up-to-date information. Administrative barriers in sharing disaster-related mortality information between key state partners may exist in locales that have decentralized ME or Coroner systems or where offices of vital statistics are not part of the health department. As a result, public health's role and ability to conduct disaster-related mortality surveillance may be hindered. To address this challenge, it is important to explore the type and format of mortality information that is readily accessible to public health for surveillance purposes during a response. Examples include daily aggregate mortality reports, access to ME line list, and preliminary death certificate data in vital statistics database. A data flow chart may help surveillance planners visualize how and by whom mortality data are captured and reported, help identify where barriers exist, and determine how data could be shared.

Large-scale disasters with mass fatalities present unique challenges to mortality reporting. Mass fatalities are inherently complex, having increased numbers of decedents, bereaved families, response partners, and media inquiries. Reporting during mass fatalities could be different if a state's electronic mass fatality database or a DMORT is deployed. In these situations, it is often unclear how mortality data will be reported for public health surveillance. A mass fatality plan should clearly describe in detail how mortality data would be shared with public health during the response. In particular, when an electronic system is to be used, it should be interoperable with the state's Electronic Death Registration System (EDRS) to facilitate rapid reporting. EDRS provides a secure method for electronically creating, updating, and certifying death certificates in jurisdictions.

Challenge: Capturing Disaster-Related Deaths in a Timely Manner

Timely disaster-related mortality surveillance provides valuable information during all phases of a disaster: response, recovery, mitigation, and preparedness. During the response, public health uses mortality data to identify ongoing hazards (e.g., carbon monoxide poisoning), provide situational awareness, allocate resources, control rumors, and target messaging.

Death Certificates and Alternative Data Sources

Death certificates are the key source of mortality data for disaster-related surveillance. Death certificate access may be delayed during protracted recovery operations with large number of decedents. When accessing or receiving death certificate information is delayed, other data sources, such as ME's or Coroner's line lists, can be used (Table 5.1). These alternate data sources may offer timelier data than death certificates and provide more detailed circumstances surrounding the death, confirmation of the disaster-relatedness of the death, and elucidate the economic burden in terms of health care and federal resource expenditure (Table 5.1). A retrospective review of data sources after a recent event could examine the usefulness of the available sources in terms of the timeliness of reporting the deaths and data quality of the alternative source.

In general, during the immediate response phase the local or state ME's or Coroner's office reports deaths through the emergency operations center and to the media. The American Red Cross (Red Cross) often receives death information within days of the death from the local certifier (i.e., ME, Coroner) so that they can begin their condolence services to families of the decedents. As the incident moves into the recovery phase, the local or state ME or Coroner's office would still provide official mortality counts and by this time, the death certificates may be entered into the EDRS. All data sources in Table 5.1 could be used to verify the final death count and to conduct research during the end of the recovery or beginning of preparedness phase. An important consideration is delayed deaths seen in the health care system. Often, deaths occurring weeks to months after the initial disaster are not captured. For example, if a victim was hospitalized for disaster-related injuries for a few weeks prior to death, the death certificate may not be coded as disaster-related. After a disaster, officials should consider a review of death information captured, used, and reported by the various response partners (i.e., Medical Examiner/Coroner, FEMA, Red Cross, and NWS) to ensure the accurate recording of the type and name of the disaster (i.e., Hurricane Ike) on the deaths certificate(s) of cases considered related to the event. This review could be part of the after action meeting(s) or "hot wash" often conducted by emergency management after disasters.

In conclusion, the challenges to defining and reporting timely mortality information are critical to address when developing disaster-related surveillance plans.

SURVEILLANCE CASE STUDIES

The following case studies illustrate the disaster-related mortality surveillance challenges and offer suggested solutions.

Hurricane Ike, Texas, 2008

Hurricane Ike (Ike), a Category 2 hurricane, made landfall on September 13, 2008, along the Galveston, Texas coast with a tremendous storm surge that led to the evacuation of nearly two million residents. The Texas Department of State Health Services has an active disaster-related mortality surveillance protocol integrated into their state incident management system. This active surveillance system is an ad hoc system with the objective of identifying and tracking deaths after a disaster. Four days after Ike, three state public health staff were assigned to actively track deaths daily until October 13, 2008. Staff emailed protocol materials to the ME and JP offices in the 44 impacted counties, including mortality surveillance report forms, definitions, and guidelines (Zane et al., 2011). ME/JPs faxed completed forms to the state where they were collated and the results were reported to their emergency operations center. State public health staff searched online news media reports for deaths that occurred before surveillance started, and ensured information about those deaths was captured. An evaluation of this surveillance system found that the system provided data useful for situational awareness and addressing media and missing person inquiries during the response phase (Choudhary et al., 2012). In addition, the evaluation also found that official death certificates underreported the number of Ike-related deaths (n = 4) compared with cases reported by the active surveillance system (n = 75) (Choudhary et al., 2012). A review of deaths captured by the Red Cross also underreported disaster-related deaths (n = 38) when compared with the active surveillance system (Fagua et al., 2013). This example highlights the utility and improved accuracy of reporting deaths attributed to the disaster when uniform case definitions and reporting protocols are in place.

Tornadoes, Southeast United States, April 2011

Among its many roles during a disaster, the Red Cross provides condolence services (e.g., grief counseling, assessment of immediate needs) to families whose loved ones were killed in a disaster. During April 25–28, 2011, a massive storm produced 351 tornadoes that killed 338 people across five southeastern states with the largest number of deaths (n = 247) in Alabama (CDC, 2013). To identify disaster-related deaths for condolence visits, Red Cross volunteers searched national and local media outlets for reports of deaths related to the tornadoes and then confirmed the death was disaster-related with the local MEs and Coroners. Once verified, volunteers visited the decedent's next of kin to offer condolence services. Condolence visits were typically done face-to-face and used a semistructured interview format to gather information about the disaster impact and the family needs. During these next of kin interviews, Red Cross completed a mortality surveillance form that captured basic demographics, location of injury, cause of the injury, and circumstances surrounding the death. The circumstances captured on the Red Cross' form provided more detailed information than is reported on the death certificates (Chiu et al., 2013). For example, the form captured whether or not the decedents heard a tornado warning. This highlights the utility of an alternative data source to provide timely information on a multistate disaster, which assisted in the completeness of reporting deaths associated with these tornadoes (Chiu et al., 2013).

Superstorm Sandy, New York City, 2012

New York City's (NYC) Bureau of Vital Statistics (Vital Statistics) used their EDRS to conduct mortality surveillance during Superstorm Sandy (Sandy). Two days before landfall, following a preestablished contingency plan, Vital Statistics identified staff and alternative work sites to ensure that reporting of disaster-related deaths would continue throughout the response phase. Staff continued to register deaths remotely after being evacuated from their offices (Howland et al., 2015). Immediately following the storm, Vital Statistics received ongoing notification from the Office of the Chief Medical Examiner about all fatal injuries that were directly related to the environmental forces of storm (e.g., storm surge) or that were direct consequences of the forces (e.g., structural collapse). To identify additional direct and indirect deaths, Vital Statistics also searched preliminary death certificates in EDRS for disaster-specific terms (e.g., "storm surge") and scanned daily media reports.

These efforts yielded additional potential cases, but on review the cases were deemed not to be related to Sandy. Reports on mortality data from EDRS on the leading causes of death provided timely data to response agencies. Analyses of EDRS data found excesses in the daily and 3-day averages of all-cause mortality after Sandy when compared with previous 2 years. This example highlights the utility of using an existing electronic system with an emergency contingency plan allowing for timely epidemiologic analysis of the Sandy-related deaths in NYC.

References

Berko, J., Ingram, D., Saha, S., & Parker, J. (2014). *Deaths attributed to heat, cold, and other weather events in the United States, 2006–2010*. National Health Statistics Reports; 76. Hyattsville, MD: National Center for Health Statistics.

Brunkard, J., Namulanda, G., & Ratard, R. (2008). Hurricane Katrina deaths, Louisiana, 2005. *Disaster Medicine and Public Health Preparedness, 2*, 215–223.

CDC. (2006). Mortality associated with hurricane Katrina — Florida and Alabama, August–October 2005. *Mortality Morbidity Weekly Review, 55*, 239–242.

CDC. (2012). Tornado-related fatalities — five states, Southeastern United States, April 25–28, 2011. *Mortality Morbidity Weekly Review, 61*, 529–533.

CDC. (2013). Deaths associated with hurricane Sandy — October–November 2012. *Mortality Morbidity Weekly Review, 62*, 393–397.

Centers for Disease Control and Prevention (CDC). (2016). *A primer for understanding the principles and practices of disaster surveillance in the United States*. Atlanta, GA: Author.

Chiu, C., Schnall, A., Mertzlufft, C., Noe, R., Wolkin, A., Spears, J., et al. (2013). Mortality from a tornado outbreak, Alabama, April 27, 2011. *American Journal of Public Health, 103*, e52–e58.

Choudhary, E., Zane, D., Beasley, C., Jones, R., Rey, A., Noe, R., et al. (2012). Evaluation of active mortality surveillance system data for monitoring hurricane-related deaths, Texas, 2008. *Prehospital and Disaster Medicine, 27*, 1–6.

Combs, D., Quenemoen, L., Parrish, R., & Davis, J. (1999). Assessing disaster-attributed mortality: development and application of a definition and classification matrix. *International Journal of Epidemiology, 28*, 1124–1129.

Fagua, N., Rey, A., Noe, R., Bayleyegn, T., Wood, A., & Zane, D. (2013). Evaluation of the American Red Cross disaster-related mortality surveillance system by using Hurricane Ike data — Texas, 2008. *Disaster Medicine and Public Health Preparedness, 7*, 13–19.

Howland, R., Li, W., Wong, H., Madsen, A., Das, T., Betancourt, F., et al. (2015). Evaluating the use of electronic death registration system for mortality surveillance during and after Hurricane Sandy, New York City, 2012. *American Journal of Public Health, 105*, e55–e62.

Ragan, P., Schulte, J., Nelson, S., & Jones, K. (2008). Mortality surveillance 2004 to 2005 Florida hurricane-related deaths. *American Journal of Forensic Medicine Pathology, 29*, 148–153.

Zane, D., Bayleyegn, T., Hellsten, J., Beal, R., Beasley, C., Haywood, T., et al. (2011). Tracking deaths related to hurricane Ike, Texas, 2008. *Disaster Medicine and Public Health Preparedness, 5*, 23–28.

CHAPTER

6

Methods: Study Designs in Disaster Epidemiology

Latasha A. Allen[1], *Jennifer A. Horney*[2]

[1]**Office of the Assistant Secretary for Preparedness and Response (ASPR), Office of Emergency Management (OEM), Washington, DC, United States;** [2]**Texas A&M University, College Station, TX, United States**

INTRODUCTION

In general, epidemiology is defined as the systematic study of the natural history of disease within populations (Institute of Medicine, 2003). In disaster setting, adverse health impacts on a population are typically caused by an acute disruption, such as a tornado or a flash flood, which leads to excess morbidity or mortality. However, more chronic environmental disruptions, such as drought and climate change, may also impact the natural history of disease within populations. Regardless, in response to natural disasters, disaster epidemiologists focus on the effects of weather events and natural phenomena to measure and describe resultant health effects, with the goals of efficiently matching resources with needs for further prevention of adverse events (Binder & Sanderson, 1987) or for the formulation of policy (Thorpe et al., 2015). Man-made or intentional disasters may present a more complex situation, requiring epidemiologists to work with law enforcement, regulatory bodies, and other agencies to design studies that can measure associations and causalities in a way that can be presented in a court of law or support the provision of behavioral health services or economic compensation to those affected (Gould, Teich, Pemberton, Pierannunzi, Larson, 2015).

Disaster epidemiologists focus on the public health impacts of natural, human-caused, or intentional disasters such as terrorism, as well as the factors that contribute to those impacts. The overall objective of a disaster epidemiology study is to assess the needs of disaster-affected populations, matching available resources to needs, preventing further adverse health effects, evaluating program effectiveness, and planning for contingencies (Noji, 1995, 1996). It is important that disaster epidemiology studies explicitly link the measurement and analysis of adverse health effects to an immediate decision-making process. Disaster

Disaster Epidemiology
http://dx.doi.org/10.1016/B978-0-12-809318-4.00008-3

epidemiologists must design studies that can quickly ascertain potential causal relationships between exposures and health outcomes to quantify who are suffering, why, how this varies between disasters, and, specifically for humanitarian/international disaster relief, the timing of these effects (Seaman, 1990).

Epidemiologists can design studies to be implemented within each phase of a disaster, collecting data on the various factors within each phase and how they contribute to the morbidity and mortality from disasters. This is similar to the methods used in outbreaks or applied epidemiology. The preimpact phase, the period of time between the recognition of an imminent hazard, and the occurrence of damage to the environment and population, is most important for planning and prevention of injuries (Binder & Sanderson, 1987). Epidemiologic studies conducted in the preimpact phase have the most potential to reduce public health effects of disasters by identifying at-risk populations, conducting vulnerability analyses, and implementing interventions to reduce deaths and postimpact medical care needs (Noji, 1992). For example, public health authorities, recognizing that the preimpact phase will bring worry, denial, and stress to some residents depending on their community, cultural, and family expectations, may focus messages on the importance of taking protective actions such as evacuation (Burkle, 1996). The impact phase of a disaster is the point when there is a sudden release of energy that results in direct injury to a population or disruption of the environment (Binder & Sanderson, 1987). In the impact phase, public health focuses on the potential effectiveness of barriers between the disaster and the population at risk. For example, a case-control study of deaths and injuries resulting from powerful tornados that impacted Oklahoma in May of 1999 showed that being in a mobile home increased the risk of death by 35% and major injury by 12% (Daley et al., 2005). Similarly, a case-control study of a 2003 heat wave that affected Europe found that elderly residents with chronic conditions and mobility problems were at greater risk as they could not easily adjust their routines to find relief from the heat (Vandentorren et al., 2006). The postimpact phase of a disaster is described at the point when the energy has dissipated and includes all subsequent disaster relief efforts. Public health impacts may include secondary injuries among emergency response workers and others helping with cleanup or exacerbation of chronic medical conditions due to low medication adherence or stress (Mokdad et al., 2005; The Hurricane Katrina Community Advisory Group and Kessler, 2007). At any time in the disaster, epidemiologic studies may include risk factors for certain types of disasters, the study of acute health effects, and longitudinal studies of the long-term health effects of disasters.

Disasters caused by humans or intentional man-made disasters include complex emergencies, technological disasters, and disasters that are not caused by natural hazards but that occur in human settlements such as transportation disasters, material shortages resulting from embargoes, and dam breaks (Noji, 1996). Studies of these types of disasters are similar in that they require that an affected population be defined. For example, in a complex emergency, many people may be displaced from their homes and some may become "refugees" in their own or in other countries. For example, after Hurricane Katrina made landfall near New Orleans, Louisiana, in 2005, more than 450,000 residents of the US Gulf Coast were displaced (Sastry & Gregory, 2014).

METHODS AND STUDY DESIGNS

Selecting the appropriate epidemiologic method and study design is an essential part of conducting effective studies when responding to a disaster. This allows epidemiologists to quickly provide valid and reliable results that can be actionable for planners and decision-makers responsible for preparedness, response, and recovery. These methods and study designs include:

- **Rapid needs assessments:** This is an assessment of the needs of a population immediately following a disaster or emergency event. Community Assessment for Public Health Emergency Response (CASPER) is a disaster epidemiology tool that can be utilized to conduct a rapid needs assessment. This method has been used to assess public health perceptions, estimate needs of a community, assist in planning for emergency response, and as part of the public health accreditation process. Regardless of the setting and objectives, once the decision to conduct the CASPER has been made, it should be initiated as soon as possible (CDC, 2012).
- **Health surveillance:** The systematic collection, analysis, and interpretation of health-related data to monitor the health of a population. Surveillance can be conducted after a disaster event to identify potential health risks and can be used to develop interventions to prepare, prevent, and control disease and injury (Thacker & Birkhead, 2008).
- **Epidemiological investigations:** Investigation to identify the source of an outbreak of disease or the causal pathway between an environmental exposure and a health outcome within a population (Fig. 6.1). The results of an epidemiologic study can assist in the coordination of the response to a public health emergency and help prevent future cases. The most common epidemiologic studies conducted in response to a disaster or public health emergency are case-control studies and cohort studies. Case-control studies start with identifying cases and investigating potential exposures or causes of illness and injury among a population or group of cases. Cohort studies begin with a defined event or exposure, for example, the collapse of the NYC World Trade Center (WTC) 9/11 attack (Brackbill et al., 2006; Wisnivesky et al., 2011) or the Deepwater Horizon oil spill (Gould, Teich, Pemberton, Pierannunzi, & Larson, 2015; Zock, Rodriguez-Trigo, POzo-Rodriguez, & Barbera, 2011), and then follow the exposed population over time to identify any potentially related health outcomes.

EXAMPLES OF DISASTER EPIDEMIOLOGY STUDIES

Rather than providing an overview of the epidemiologic study designs, which is available in any introductory textbook, we have chosen six different events, including intentional, man-made, and natural disasters, to explore the use of epidemiologic study designs.

THE WORLD TRADE CENTER, SEPTEMBER 11, 2001

The attack on the WTC's "Twin Towers" on September 11, 2001, was a dramatic event for the United States, which led to many public and environmental health concerns among

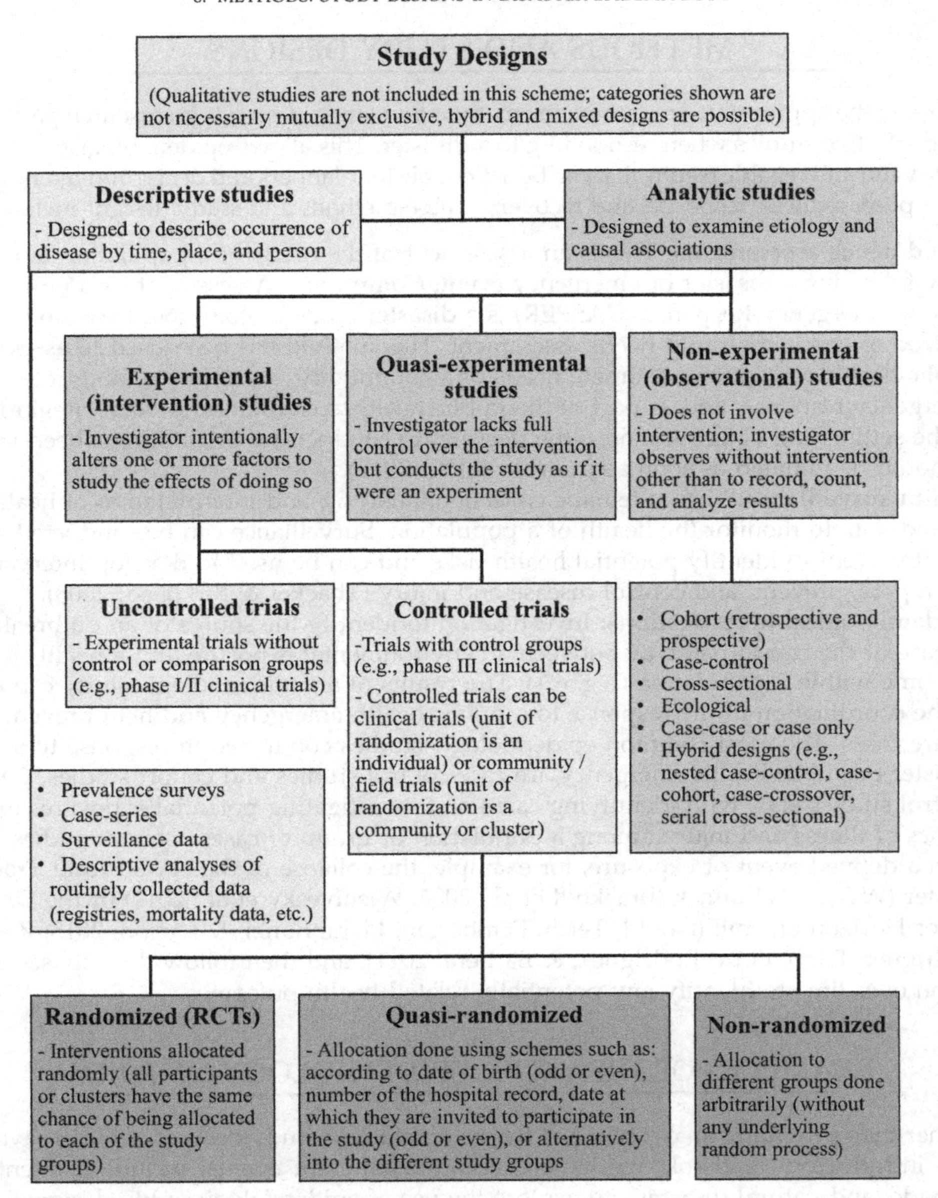

FIGURE 6.1 Epidemiologic study designs. *Adapted with permissions from Pai, M., & Filion, K. (n.d.). Classification of study designs (Version 8). Retrieved from http://www.teachepi.org/documents/courses/Classification%20Design.pdf.*

survivors, responders, and witnesses of the tragedy. One example of this was the exposure of New York City residents to caustic dust particles following the buildings collapse (Svendsen et al., 2012). To address this concern, the WTC Health Registry, an epidemiological cohort study, was created after 9/11 to assist with the public health response (Brackbill et al., 2006). This registry was available for voluntary enrollment to people who were in the area of the WTC disaster (City of New York, 2014). The registry was used to monitor both mental and physical health, and showed that in the years after 9/11, survivors of the WTC attack had significant increases in both physical and mental health issues (Farfel et al., 2008). Another cohort, including WTC rescue and recovery workers, was recruited and followed to document the health impacts on rescue and recovery workers (Wisnivesky et al., 2011). The sharing of findings from these cohort studies with first responders and others in the emergency and disaster management community could be critically important to preventing some of the health impacts in future responses (Brackbill et al., 2006).

HURRICANE KATRINA AND SUPERSTORM SANDY

Hurricane Katrina and Superstorm Sandy were both major domestic natural disasters; however, their impacts on population health were very different. Hurricane Katrina made landfall on August 29, 2005, along the US Gulf Coast, a region very familiar with the threat of hurricanes. Many long-term residents of the region, having experienced other storms that were near misses or not as strong as predicted, including Hurricane Ivan only the year before Katrina, chose to ride out storm. The result was both a natural and a man-made disaster. As a result of Katrina, a Category 3 hurricane, the levee system that protected the City of New Orleans failed.

After Katrina, the Centers for Disease Control and Prevention (CDC) assisted the Mississippi Department of Health (MDH) in conducting a cross-sectional rapid community needs assessment in Hancock County, one of the areas most affected by Katrina. The purpose of the assessment was to identify the public health needs of the community and estimate the effect of the hurricane on households to assist response and recovery activities (McNeil et al., 2006). A one-page survey was used to interview households to collect information about unmet basic needs (e.g., food, shelter, prescription medications, and utilities), as well as illnesses and injuries that may have been related to the storm. Participating households were randomly selected using geographical information systems (GIS) tools, with global positioning systems (GPS) used to direct survey teams to the location of the selected households. Information was rapidly provided to both the MDH and Mississippi Emergency Management and evaluated to determine the continued need for relief services. The needs of those who had evacuated were also assessed. For example, Rogers et al. (2006) conducted a rapid assessment of a sample of approximately 12,700 Hurricane Katrina evacuees who were housed in evacuation centers in San Antonio, Texas.

In contrast, Superstorm Sandy impacted the coast of the Northeastern United States, a region that was less familiar with the potential impacts of a major hurricane. This unfamiliarity with tropical storms having the strength of Sandy resulted in an increase in preventable deaths and injuries. For example, in the New York City metropolitan area, 97 people died in the storm, thousands were displaced, and two major hospitals required evacuation (Abramson & Redlener, 2012). In southern New Jersey, federal and state public health professionals conducting regular postdisaster surveillance activities discovered an increase in reported exposure to carbon monoxide (CO) to poison control centers. After active case finding, 263 cases of CO exposure were identified in eight states impacted by Superstorm Sandy

(80, New York; 61, New Jersey; 44, Connecticut; 39, Pennsylvania; 27, West Virginia; 8, Virginia; 3, Maryland; and 1, Delaware) (Clower et al., 2012). In New York City, a retrospective review of CO exposure data from two surveillance systems, the New York City Poison Control Center (NYCPCC) and New York City's Syndromic Surveillance Unit, was conducted to collect CO exposure reports. Data from these systems were compared with CO exposure data from identical time periods between 2008 and 2011 and results showed that there was a significant increase in 2012 when compared with CO exposure reports from the preceding 4 years (Chen et al., 2013). The results of these longitudinal studies could benefit immediate response activities, as well as future emergency planning and preparedness efforts.

Across the area, hospitals, nursing homes, and assisted living facilities were evacuated after Hurricane Sandy made landfall, resulting in major public health challenges (Powell, Hanfling, & Gostin, 2012). For example, Bellevue Hospital evacuated patients in the immediate aftermath of the storm when the backup power source failed. The medical and public health challenges after Hurricane Sandy were so significant that they were referred to as "medical humanitarianism," including volunteer doctors working in mobile medical units and nurses and clinicians who staffed the large number of shelters (Abramson & Redlener, 2012).

Although many types of quantitative disaster epidemiology studies were conducted after Sandy, a qualitative study was used to identify lessons learned within the health-care sector. To collect qualitative data, public meetings or forums were hosted by the Institute of Medicine (IOM) and New York Academy of Medicine (NYAM). At these forums, participants identified a range of unmet needs related to public health and health care post-Sandy, including real-time data pertaining to hospital and shelter capacity, reentry/repatriation regulations and procedures, information sharing, particularly among the Federal Emergency Management Agency (FEMA), hospitals, and shelters, and guidelines regarding rules and waivers for occupancy, patient confidentiality, and admitting privileges after evacuations (Alancantara, 2012).

DEEPWATER HORIZON OIL SPILL

Less than 5 years after Hurricane Katrina, the US Gulf Coast was impacted by a major man-made disaster. On April 20, 2010, the Deepwater Horizon oil rig exploded, releasing approximately 5 billion barrels of crude oil into the Gulf of Mexico, causing an ecological and public health disaster. More than 100,000 responders took part in the ongoing response and cleanup, and several epidemiologic studies were implemented to assess the short- and long-term health effects of the spill (Zock et al., 2011). In a region that was slowly recovering from one of the worst natural disasters to ever impact the United States, the mental health of both residents and responders was of concern. Epidemiologists from CDC conducted a mental health needs assessment in coastal areas of Alabama and Mississippi to assess mental health status. Results showed that 31.5% of residents reported symptoms consistent with an anxiety disorder, 22.8% reported ≥14 mentally unhealthy days within the past 30 days, 35.7% of households reported decreased income since the oil spill, and 38.2% of households reported having been exposed to oil (Buttke, Bayleyegn et al., 2012). When compared with baseline data from the 2006 and 2009 Behavioral Risk Factor Surveillance System, residents of areas affected by the oil spill had higher proportions of negative mental health parameters, suggesting that the public health response should focus on mental health services within these communities (Buttke, Bayleyegn et al., 2012). In a follow-up study conducted 1 year later, the percentage of reported mental health problems still exceeded baseline levels but had fallen since 2010 (Buttke, Schnall et al., 2012). These findings suggest that emergency responders and others involved in cleanup and recovery should receive at least basic mental health training so they know how to recognize conditions such as posttraumatic stress disorder that may require a referral for mental health services.

TYPHOON HAIYAN, PHILIPPINES

In November, 2013, the eastern Philippines was devastated by Typhoon Haiyan, which impacted over 11 million people. Various international agencies and organizations participated in humanitarian relief efforts to assist the regions facing this catastrophe. Epidemiology experts from CDC were deployed to assist with the postdisaster response and collaborate with the World Health Organization (WHO) and the Philippines Department of Health (DOH) as part of the emergency management team to strengthen emergency disease surveillance systems. In part because regular surveillance activities were suspended in the most affected region, public health relief efforts focused on responding to vaccine shortages, an outbreak of measles, and preparing for potential infectious disease outbreaks including cholera, other diarrheal diseases, and dengue fever (Levy, Iyengar, Nakao, Easley, & Woodring, 2014). In addition, a case-control study was conducted to determine the cause of an outbreak of gastrointestinal illness, which was found to be associated with contamination of a municipal water system (Ventura, Muhi, de los Reyes, Sucaldito, & Tayag, 2015). About 16 months after Typhoon Haiyan, a study of the status of the Philippine Integrated Disease Surveillance and Response, the system the DOH uses to conduct surveillance for 25 diseases and syndromes that have the potential to cause outbreaks, was conducted (Gallardo et al. 2015). All epidemiology and surveillance units reported that they were performing all core functions of surveillance; however, laboratory capacity was still reduced.

HAITI EARTHQUAKE, 2010

On January 12, 2010, a 7.0 earthquake with the epicenter located in the capital city of Port-au-Prince, Haiti devastated the country and killed more than 200,000 residents. Humanitarian public health efforts were quickly coordinated by the Haiti Ministry of Public Health and Population (MSPP), the Pan American Health Organization (PAHO), CDC, and other national and international agencies. Within 2 weeks of the earthquake, the National Sentinel Site Surveillance (NSSS) System was launched (Magloire et al., 2010). Data collected through the system were used to respond to rumors of disease outbreaks and provide explanations for disease concerns. Data from the NSSS were also linked with data from the Internally Displaced Persons Surveillance System, which was implemented to capture data on the status of the displaced population. Both systems were used to routinely collect postdisaster data so that disease trends could be used to identify gaps in disaster response and support decision-making around the rebuilding of health-care infrastructure.

CONCLUSION

This chapter reviewed various types of emergencies and disasters where officials utilized disaster epidemiologic methods, tools, and study designs to assist with postimpact response. As seen in the examples discussed, which included both domestic and international events, as well as man-made and natural disasters, epidemiologic tools were mostly used to assess the needs of the affected communities postdisaster. In the future, disaster epidemiologists should seek a more substantial role during the planning and preimpact phases of disasters. For example, data gathered as part of preparedness studies conducted in the preimpact phase can be shared with emergency managers to help prepare for the public health needs of affected communities and shape planning for future response and recovery efforts. These types of preimpact data can be used to quantify the potential consequences of future emergencies and disasters, such as communicable disease outbreaks or mental health needs, to ensure that the needs of impacted populations can be met (CDC, 2012). When combined with other types of data, such as ecological, meteorological, spatial, or social vulnerability data, preimpact disaster epidemiology studies can be used to predict and plan for the potential social impacts of a disaster and spatially and temporally assess where and when different resources will be needed after various types of emergencies and disasters. This type of predictive analysis can be very important in planning for low probability/high consequence events and complex emergencies in both the domestic and international setting by suggesting ways to increase the efficiency and cost effectiveness of response and relief efforts. As the field of disaster epidemiology continues to develop, one important focus should be identifying ways in which epidemiological studies can be beneficial in all phases of disaster, including preparedness, response, recovery, and mitigation.

Disclaimer

The opinions, findings, and conclusions in this text are those of the author(s) and do not necessarily represent the views of the U.S. Department of Health and Human Services or its components.

References

Abramson, D. M., & Redlener, I. E. (2012). Hurricane Sandy: Lessons learned, again. *Disaster Medicine and Public Health Preparedness, 6*(4), 328–329.

Alancantara, R. (2012). *Identifying disaster medical and public health research priorities: Data needs arising in response to Hurricane Sandy*. New York, NY: New York Academy of Medicine.

Binder, S., & Sanderson, L. M. (1987). The role of the epidemiologist in natural disasters. *Annals of Emergency Medicine, 16*(9), 1081–1084.

Brackbill, R. M., Thorpe, L. E., DiGrande, L., Perrin, M., Sapp, J. H., Wu, D., et al. (2006). Surveillance for World Trade Center disaster health effects among survivors of collapsed and damaged buildings. *Morbidity and Mortality Weekly Report, 55*(SS02), 1–18.

Burkle, F. M. (1996). Acute-phase mental health consequences of disasters: Implications for triage and emergency medical services. *Annals of Emergency Medicine, 28*(2), 119–128.

Buttke, D., Vagi, S., Bayleyegn, T., Sircar, K., Strine, T., Morrison, M., et al. (2012a). Mental health needs assessment after the Gulf Coast oil spill—Alabama and Mississippi, 2010. *Prehospital and Disaster Medicine, 27*(5), 401–408.

Buttke, D., Vagi, S., Schnall, A., Bayleyegn, T., Morrison, M., Allen, M., et al. (2012b). Community assessment for public health emergency response (CASPER) one year following the Gulf Coast oil spill: Alabama and Mississippi, 2011. *Prehospital and Disaster Medicine, 27*(6), 496–502.

CDC. (2012). *Community Assessment for Public Health Emergency Response (CASPER) toolkit* (2nd ed.). Atlanta, GA: CDC.

Chen, B. C., Shawn, L. K., Connors, N. J., Wheeler, K., Hoffman, R. S., Matte, T. D., et al. (2013). Carbon monoxide exposures in New York city following Hurricane Sandy in 2012. *Clinical Toxicology, 51*(9), 879–885.

City of New York. (2014). *World Trade Center Health Registry*. New York, NY: NYC 9/11 Health. Retrieved from http://www1.nyc.gov/site/911health/about/wtc-health-registry.page.

Clower, J., Henretig, F., Trella, J., Hoffman, R., Wheeler, K., Maxted, A., et al. (2012). Notes from the field: Carbon monoxide exposures reported to poison centers and related to Hurricane Sandy-Northeastern United States, 2012. *Morbidity and Mortality Weekly Report, 61*(44), 904–905.

Daley, W. R., Brown, S., Archer, P., Kruger, E., Jordan, F., Batts, D., et al. (2005). Risk of tornado-related death and injury in Oklahoma, May 3, 1999. *American Journal of Epidemiology, 161*(12), 1144–1150.

Farfel, M., DiGrande, L., Brackbill, R., Prann, A., Cone, J., Friedman, S., et al. (2008). An overview of 9/11 experiences and respiratory and mental health conditions among World Trade Center Health Registry enrollees. *Journal of Urban Health, 85*(6), 880–909.

Gallardo, F. D. L., de los Reyes, V. C., Sucaldito, M. N., Ligon-Imperio, L., Peñas, J., Rebato, N., et al. (2015). An assessment of the case notification system 16 months after Typhoon Haiyan in Region 8, the Philippines. *Western Pacific Surveillance and Response Journal, 6*(Suppl. 1), 71–75.

Gould, D. W., Teich, J. L., Pemberton, M. R., Pierannunzi, C., & Larson, S. (2015). Behavioral health in the Gulf Coast region following the Deepwater Horizon oil spill: Findings from two federal surveys. *Journal of Behavioral Health Services and Research, 42*(1), 6–22.

Institute of Medicine. (2003). *The future of the Public's health in the 21st Century*. Washington, DC: The National Academy Press.

Levy, B., Iyengar, P., Nakao, J., Easley, S. J., & Woodring, J. (2014). Emergency response and recovery: Typhoon Haiyan, Philippines. *Updates from the Field*. Spring(14):7.

Magloire, R., Mung, K., Harris, S., Bernard, Y., Jean-Louis, R., Niclas, H., et al. (2010). Launching a national surveillance system after an earthquake- Haiti, 2010. *Morbidity and Mortality Weekly Report, 59*(30), 933–938.

McNeil, M., Goddard, J., Henderson, A., Phelan, M., Davis, S., Wolkin, A., et al. (2006). Rapid community needs assessment after Hurricane Katrina - Hancock County, Mississippi, September 14–15, 2005. *MMWR. Morbidity and Mortality Weekly Report, 55*(9), 234–236.

Mokdad, A. H., Mensah, G. A., Posner, S. F., Reed, E., Simoes, E. J., & Engelgau, M. M. (2005). When chronic conditions become acute: Prevention and control of chronic diseases and adverse health outcomes during natural disasters. *Preventing Chronic Disease, 2*(Suppl. 1), A04.

Noji, E. K. (1992). Disaster epidemiology: Challenges for public health action. *Journal of Public Health Policy, 13*(3), 332–340.

Noji, E. K. (1995). Disaster epidemiology and disease monitoring. *Journal of Medical Systems, 19*(2), 171–174.

Noji, E. K. (1996). Disaster epidemiology. *American Journal of Disaster Medicine, 14*(2), 289–300.

Pai, M., & Filion, K. (n.d.). Classification of study designs (Version 8). Retrieved from http://www.teachepi.org/documents/courses/Classification%20Design.pdf.

Powell, T., Hanfling, D., & Gostin, L. O. (November 2012). Emergency preparedness and public health: The lessons of Hurricane Sandy. *Journal of the American Medical Association, 308*(24), 2569–2570.

Rogers, N., Guerra, F., Suchdev, P., Chapman, A., Plotinsky, R., Jhung, M., et al. (2006). Rapid assessment of health needs and resettlement plans among Hurricane Katrina evacuees – San Antonio, TX, September 2005. *Morbidity and Mortality Weekly Report, 55*(9), 242–244.

Sastry, N., & Gregory, J. (2014). The location of displaced New Orleans residents in the year after Hurricane Katrina. *Demography, 51*(3), 753–775. http://dx.doi.org/10.1007/s13524-014-0284-y.

Seaman, J. (1990). Disaster epidemiology: Or why most international disaster relief is ineffective. *Injury, 21*(1), 5–8.

Svendsen, E. R., Runkle, J. R., Dhara, V. R., Lin, S., Naboka, M., Mousseau, T. A., et al. (2012). Epidemiologic methods lessons learned from environmental public health disasters: Chernobyl, the World Trade Center, Bhopal, and Graniteville, South Carolina. *International Journal of Environmental Research and Public Health, 9*(8), 2894–2909.

Thacker, S. B., & Birkhead, G. S. (2008). Surveillance [Chapter 3]. In M. B. Gregg (Ed.), *Field epidemiology* (3rd ed.). New York, NY: Oxford University Press.

The Hurricane Katrina Community Advisory Group, & Kessler, R. C. (2007). Hurricane Katrina's impact on the care of survivors with chronic medical conditions. *Journal of General Internal Medicine, 22*(9), 1225–1230. http://dx.doi.org/10.1007/s11606-007-0294-1.

Thorpe, L. E., Assari, S., Deppen, S., Glied, S., Lurie, N., Mauer, M. P., et al. (2015). The role of epidemiology in disaster response policy development. *Annals of Epidemiology, 25*(5), 377–386.

Vandentorren, S., Bretin, P., Zeghnoun, A., Mandereau-Bruno, L., Croisier, A., Cochet, C., et al. (2006). August 2003 heat wave in France: Risk factors for death of elderly people living at home. *The European Journal of Public Health, 16*(6), 583–591.

Ventura, R. J., Muhi, E., de los Reyes, V. C., Sucaldito, M. N., & Tayag, E. (2015). A community-based gastroenteritis outbreak after Typhoon Haiyan, Leyte, Philippines, 2013. *Western Pacific Surveillance and Response Journal, 6*(1), 1–6. http://dx.doi.org/10.2471/WPSAR.2014.5.1.010.

Wisnivesky, J. P., Teitelbaum, S. L., Todd, A. C., Boffetta, P., Crane, M., Crowley, L., et al. (2011). Persistence of multiple illnesses in World Trade Center rescue and recovery workers: A cohort study. *Lancet, 378*(9794), 888–897.

Zock, J.-P., Rodriguez-Trigo, G., POzo-Rodriguez, F., & Barbera, A. J. (2011). Health effects of oil spills: Lessons from the Prestige. *American Journal of Respiratory and Critical Care Medicine, 184*(10), 1094–1096.

Further Reading

Novick, L. F. (2005). Epidemiologic approaches to disaster: Reducing our vulnerability. *American Journal of Epidemiology, 162*(1), 1–2.

CHAPTER

7

Applications: Community Assessment for Public Health Emergency Response

Amy H. Schnall, Amy Wolkin, Tesfaye M. Bayleyegn
Centers for Disease Control and Prevention, Atlanta, GA, United States

INTRODUCTION

With every disaster, there is a need for public health data to make decisions, allocate resources, and provide situational awareness. Epidemiologic methods, including rapid needs assessments, can provide reliable and actionable data (Malilay et al., 2014). A Rapid Needs Assessment (RNA) is a systematic process of information collection and analysis regarding the type, depth, and scope of a problem (United States Agency for International Development, 2014). RNAs can provide evidence to inform and enhance response capability within the public health infrastructure through quick and effective action (Pan American Health Organization, 2014). RNAs have been recognized as a useful method for determining the immediate needs of communities after disasters (Malilay, Flanders, & Brogan, 1996). An RNA should be timely and methodologically sound so it can provide key information to identify health problems and establish priorities for decision-makers. Poorly conducted assessments are likely to lead to poor planning decisions, an inadequate response, and, often, consequences beyond the disaster phase into recovery efforts (International Federation of Red Cross and Red Crescent Societies, n.d.).

There are many types of RNAs, varying in a wide range of topics as well as methodologies (e.g., cluster sampling, convenience sampling, and purposive sampling). For example, the International Federation of Red Cross and Red Crescent Societies (IFRC) recommends observations, discussions with key informants, and group interviews for RNAs in emergencies (IFRC, 2008); the Federal Emergency Management Agency (FEMA) conducts an RNA through a "windshield survey" or aerial overflight to determine immediate resource needs of the affected area (FEMA, 2001); and the Centers for Disease Control and Prevention (CDC) recommends a two-stage cluster methodology to assess the public health needs of a community (Bayleyegn et al., 2012). This chapter will describe the CDC's methodology.

Disaster Epidemiology
http://dx.doi.org/10.1016/B978-0-12-809318-4.00009-5

BACKGROUND AND HISTORY

CDC's Community Assessment for Public Health Emergency Response (CASPER) is modeled after the World Health Organization's (WHO) Expanded Programme on Immunization (EPI) survey technique for estimating vaccine coverage (WHO, 2005). In the 1970s, EPI implemented the two-stage design as the preferred method of rapid estimation of vaccine coverage in Africa. This design utilizes a sampling method where in the first stage, a set number of clusters (i.e., mutually exclusive groups in a population), are selected (e.g., 30), and in the second stage, a set number of persons or households are interviewed (e.g., 7) from each selected cluster. A commonly used, and recommended, two-stage cluster sampling design is 30 × 7; however, the number of clusters and households selected is determined based on the desired precision. The 30 × 7 design is aimed to achieve a desired precision of ±10% points. Meaning that, if the true prevalence was 25%, one would expect an estimate between 15% and 35% when using the 30 × 7 method (WHO, 2005). In 1985, the WHO commissioned Lemeshow and Robinson (1985) to statistically justify this EPI survey methodology. And, in 1992, the EPI methodology was adopted for disaster response following the impact of Hurricane Andrew in Florida (Malilay et al., 1996). Due to the increased use of the two-stage 30 × 7 design for disaster response following Hurricane Andrew, the Division of Environmental Hazards and Health Effects, Health Studies Branch (DEHHE/HSB) at the CDC published a toolkit outlining the methodology. To avoid confusion with other types of RNA, the CDC renamed this two-staged 30 × 7 survey methodology for public health assessment the CASPER.

CASPER is an epidemiologic tool designed to provide quick, reliable data to leaders and decision-makers. Gathering health and basic needs information using valid statistical methods allows public health and emergency managers to make informed decisions. CASPER data are also often used to efficiently allocate limited resources, provide valid information to the news media to dispel rumors, support the funding of projects, and for future planning purposes. For example, during the 2009 Kentucky ice storms, CDC and the Kentucky Department for Public Health (KDPH) conducted a CASPER and identified the community's need for additional supplies of supplemental oxygen (CDC, 2009). KDPH also used CASPER data to determine the need to disseminate targeted public health messages on the dangers of carbon monoxide, educate the community on the safe use of generators, and modify future emergency plans to account for gaps such as pet-friendly shelters and early event communication messaging.

While CASPER was originally designed to provide information during disaster response, CASPER can also be used whenever population-representative data are needed throughout the disaster life cycle (preparedness, response, recovery, mitigation). CASPERs conducted during the preparedness phase of the disaster life cycle may focus on evacuation planning and household readiness, as well as communications, so that local emergency plans can be tailored to the specific needs and requirements of the community. Response CASPERs often focus on the immediate needs and general health status of the community. Specific questions on behaviors and actions taken during, or prior to, the disaster impact, such as protective actions, may also be included. During the disaster recovery phase, CASPERs can be used to assess the long-term and ongoing needs of the community and can also be used to evaluate response efforts or programs. CASPER has also proven useful for nondisaster purposes. An increasing number of CASPERs are being conducted throughout the United States to quickly obtain population-representative data. For example, public health departments have used CASPER to find out household-level information about chronic respiratory conditions, knowledge of emerging infectious diseases such as Zika virus and H1N1, and community awareness, opinions, and concerns on subjects such as new coal gasification plants, healthy homes, and community health (CDC, 2016a).

SAMPLING FRAME AND METHODOLOGY

Regardless of the reason for conducting a CASPER, there are several key steps to take in preparing for field deployment. The leadership team must determine the partners to be involved, decide roles and responsibilities, define the assessment area (sampling frame), develop the questionnaire and relevant forms (or refine existing documents), create an analysis and reporting plan, and secure any funding needed. All partners need to agree on objectives and then need to determine whether the CASPER methodology is appropriate to meet the objectives. For example, if the objective is to assess the needs of pregnant women after a disaster, then CASPER is likely not the most appropriate method and sampling methods that target small subpopulations may be more suitable. Additional methodological considerations include number of households in and geographic size of the community. As a general guide, CASPERs are conducted in areas with a minimum of 800 households. Areas with less than 800 households typically do not have enough clusters with an adequate number of households and thus require a different (i.e., noncluster) sampling approach such as systematic or stratified sampling (Bayleyegn et al., 2012). Geographic size also plays a role in the amount of time it may take to conduct the CASPER, with very large geographic areas potentially requiring more time for data collection. The time taken to conduct a CASPER can also be affected by the disaster itself, especially when transportation or access is an issue.

The timing of the CASPER will be determined by the objectives and, if during a disaster, the safety of the area and when evacuation orders are lifted. The CASPER should be initiated as soon as logistically possible and has been conducted as early as within 24 h of disaster impact. The sampling frame is all households in a selected geographic area from which the sample is drawn and is determined by the objectives of the CASPER. It can be defined in various ways and may be based on political boundaries, by county, city, jurisdiction, or neighborhood; geographic boundaries, by key landmarks, roads, or rivers; or situationally, by a storm track, most affected area, social vulnerability, or other factor. For example, if the objective is to determine the needs of the community impacted by a flood, then flood maps will be used to determine what community(ies) should be included (Horney, MacDonald, Van Willigen, Berke, & Kaufman, 2010a, 2010b; Wallace, Poole, & Horney, 2016). Once the sampling frame is identified, the first stage of sampling is conducted. Thirty (30) clusters, often census blocks, are selected with a probability proportional to the number of households within the cluster. The recommended method for cluster selection is through the use of the CDC CASPER ArcGIS toolkit, which rapidly selects clusters based on US Census data (CDC, 2016b). ArcGIS provides flexibility in the selection of a sampling frame compared with manual selection, which is limited to counties or groups of counties, ensures the selection process is conducted according to the CDC CASPER methodology, and is more efficient than manually selecting clusters.

The second stage of sampling involves selecting seven households for interviews and is conducted systematically in the field by interview teams. Two-person interview teams are assigned clusters, provided with detailed maps, and instructed to go to every nth household (where "n" is the total number of households in the cluster divided by seven). The goal is to interview seven households per cluster. Teams should make multiple attempts (e.g., three separate visits) at each selected household before substitution of a household. Household substitution, although not ideal, ensures that enough households are interviewed to achieve statistical validity in a timely manner.

Occasionally, changes to the CASPER methodology are made. The decision to modify the methodology must be made prior to sampling. For example, if there is concern that clusters are not accessible, oversampling of clusters may be an option. Clusters should always be chosen without substitution—meaning that all clusters selected are the clusters that are assessed. For example, if you are concerned that some of the clusters may be inaccessible due to downed trees, you may select 40 rather than 30 clusters. However, you will need to visit all 40 clusters and not limit your sample to the first 30 clusters completed. While increasing the number of clusters selected in the first stage of sampling will increase your sample size (and, therefore, your power), it will not necessarily improve your response rate as it does not increase the accessibility of clusters.

CASPER methodology may also be modified in the second stage. Systematic sampling is always recommended as it facilitates interviews to be distributed throughout the cluster. However, in certain circumstances, sequential sampling may be necessary. For example, because the affected area was rural, many homes were destroyed, and the terrain was difficult to traverse after the devastating wildfires in Bastrop County, Texas in 2011, teams randomly selected a starting point within their cluster and then selected households sequentially. This allowed teams to safely and effectively move through their assigned clusters. It is always recommended to talk to a CASPER subject matter expert or statistician before modifying the sampling methodology. CDC CASPER experts can provide viable options that will fit the specific needs of the situation.

QUESTIONNAIRE AND FIELD INTERVIEWS

While in the field, teams conduct the interviews, track all households they attempt to interview, and provide, if available, public health information materials to all potential respondents and interested persons. While CASPER is not intended to provide any direct service to the community, interview teams should have referral forms to share with any households with an urgent need (e.g., supplemental oxygen, mental health services, prescription medication).

Verbal consent is sufficient for CASPER. The consent script is typically an introduction to the CASPER survey and provides the potential respondent with information about who the team is, why they are there, and how long the interview will take and explicitly requests participation. CASPER interviews typically comprise a two-page questionnaire and aim to be approximately 15–20 min in length. All questions should be directly related to the CASPER objectives. This will help keep the questionnaire brief, assure that whatever critical information needed is being collected, and ensure that all data are useful. To decrease analysis time, closed-ended questions that request only information that will satisfy the objectives are ideal. Examples of questionnaires are available from CDC and online in CASPER reports.

Data are collected in either paper format or electronic form. Both the paper and electronic forms have advantages and disadvantages. Generally, while a paper form can be more labor intensive in the data entry process, the electronic form can be labor intensive during questionnaire development and training. Due to the rapid nature of CASPER implementation, there is often not enough time to adequately develop the form electronically while accounting for potential last-minute edits in the field, uploading to the devices, and pilot testing. Table 7.1 provides considerations for implementing a questionnaire in paper versus electronic form. Interview teams typically conduct fieldwork for approximately 10 h over a 2-day period (e.g., 2–7 p.m. each day). CASPER fieldwork should be scheduled using local knowledge of the community to ensure the best times to reach people at home.

TABLE 7.1 Considerations of Using Paper Forms Versus Electronic Forms for Community Assessment for Public Health Emergency Response Data Collection

Paper Form	Electronic Form
• No technical training	• Technical training required • Potential to be awkward or slow for those teams not accustomed to the technology
• Relatively cheap supplies	• May be expensive to purchase the hardware and software • May incur costly damage in the field if broken, dropped, or water damaged
• Requires paper, pens, and clipboards in the field	• Requires data collection devices and battery chargers in the field
• No maintenance of supplies	• Requires maintenance and care of software and devices
• Can be labor intensive to enter data into database after fieldwork	• Can be labor intensive to develop electronic questionnaire prior to fieldwork
• Potential for error in manual transfer of data from paper to database	• Can provide real-time data quality checks
• Relatively slow data management processes: requires data entry after fieldwork	• Data management process is quicker; no data entry required after the fieldwork
• No limitation on the number of field teams (provided the necessary personnel are available)	• May limit the number of field teams due to availability of equipment

DATA ANALYSIS AND REPORTING

Once data are collected and entered electronically, the data are analyzed. Households selected in cluster sampling have an unequal probability of selection. To avoid biased estimates, *all data analyses should include a mathematical weight for probability of selection.* Once all data are merged into a single electronic data set, a weight variable must be added to each surveyed household by use of the formula in Fig. 7.1.

The numerator is the total number of households in the defined sampling frame and the denominator is the total number of households interviewed within a given clusters (i.e., 1–7) multiplied by the number of clusters selected (i.e., 30). If sampling has been 100% successful, meaning a survey was obtained in exactly 7 households in all 30 selected clusters, the denominator will be 7 × 30, or 210, for every household. Likely, teams will not complete all seven interviews in each cluster; therefore, the denominator will differ depending on the cluster from which the household was selected. In other words, households from the same cluster will always have the same weight, but weights may differ between clusters. Confidence intervals (e.g., 95%) should be provided to indicate the reliability of the weighted estimate. These can be calculated in any statistical program that accommodates weighted data (e.g., Epi Info 7, SAS).

To help determine the representativeness of the sample to the population within the sampling frame, response rates are calculated. All information used to calculate response rates is collected by the interview teams on the tracking form. The tracking form collects information about each household selected. For example, after a home is visited, the interviewer will record "no answer" for a house that did not have anyone home. Teams must be trained to use the tracking form to ensure accuracy of response rates. There are three separate response rates that are calculated for CASPER: the contact rate, the cooperation rate, and the completion rate (Fig. 7.2).

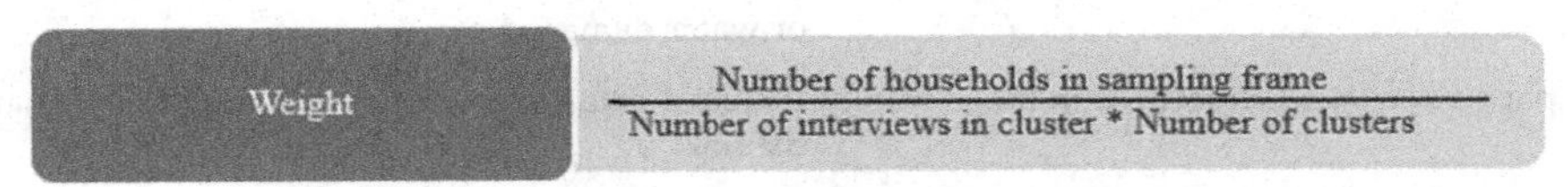

FIGURE 7.1 Formula to calculate mathematical weight.

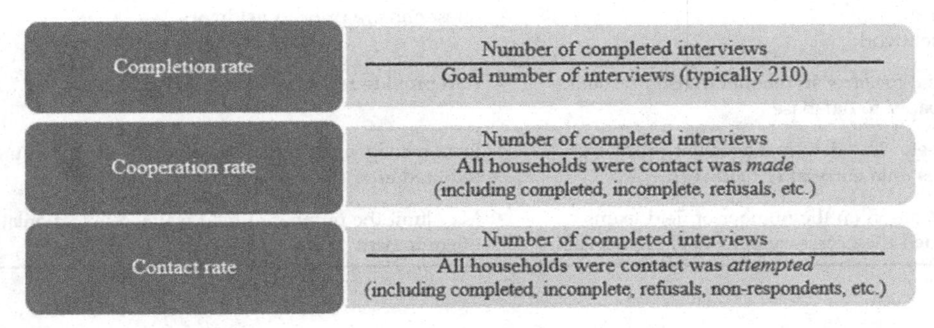

FIGURE 7.2 Calculation of Community Assessment for Public Health Emergency Response rates.

The contact rate is the proportion of households who successfully completed an interview compared to all households at which contact was attempted. Higher contact rates indicate better representativeness of the sample to the population because a larger number of the households originally selected participated in the CASPER. Lower contact rates indicate that the teams had to knock on many doors to obtain the seven interviews. As the number of households approached increases, the sample becomes less statistically representative of the population and more like a convenience sample.

The cooperation rate is the proportion of households at which the interview was successfully completed compared with those households at which contact was made (i.e., someone at the household answered the door). Cooperation rates are outcome rates among those eligible and therefore are typically higher than contact rates.

Finally, the completion rate is the proportion of households at which the interview was successfully completed compared with the goal (typically n = 210). Higher completion rates indicate that the interview teams came closer to completing 210 interviews. Completion rates below 80% (typically n = 168) result in an unacceptably low number to statistically represent the sampling frame and are not sufficient to perform weighted population estimates. Once an 80% or more completion rate is obtained, the CASPER is considered successful, and a report should be presented to public health leadership and decision-makers within 36 h. The rapid turnaround of preliminary results is essential so that leaders and decision-makers can take action. Consideration should be given to the best way to present the results in a manner to ensure timely and effective response. A more concrete, finalized, and cleaned report may be given at a later date.

The CDC CASPER toolkit and supporting webpage provide detailed guidelines on questionnaire development, methodology, sample selection, training, data collection, analysis, and report writing as well as numerous examples and templates. A list of completed CASPERs is available on the CDC webpage (CDC, 2016a).

CASPER EXAMPLE

Household Emergency Preparedness—Oakland County, Michigan

Oakland County, Michigan, located near the Great Lakes, is subject to extreme weather such as winter storms and tornadoes, as well as other emergency events such as power outages and chemical spills. The 1.2 million residents also reside approximately 50 miles from the Fermi Nuclear Power Plant. As part of Michigan's capability-based planning for public health emergency preparedness, and in conjunction with National Emergency Awareness Month, the Oakland County Health Division (OCHD), the Michigan Department of Community Health (MDCH), and the CDC conducted a CASPER in September 2012 to assess levels of household preparedness in Oakland County and to develop public health capacity for emergency preparedness and response, including radiation emergencies.

The sampling frame comprised the 575,255 households in Oakland County (U.S. Census Bureau, 2010). Adherent to the CASPER methodology, 30 clusters were selected with a probability proportional to the number of households within the cluster and seven households were selected by field interview teams systematically within each chosen cluster. Teams made three attempts at each selected household before substituting another household within the cluster. The teams determined housing type as single-detached homes or multiunit dwellings, which included duplexes, townhomes, condominiums, and apartment buildings. The two-page questionnaire included questions about the household members' general health and special needs; communication information such as the most trusted and main source of information for households; and self-reported household emergency preparedness, including emergency plans, training in emergency response, and ownership of emergency supplies.

Teams completed a total of 192 interviews (91% completion rate). The most common health conditions reported by households were hypertension/heart disease, respiratory disease, and diabetes. During a radiation emergency, the majority of households stated they would follow instructions (e.g., shelter-in-place) from an official and that the local public health department would be the most trusted source of information (Fig. 7.3). Overall, the

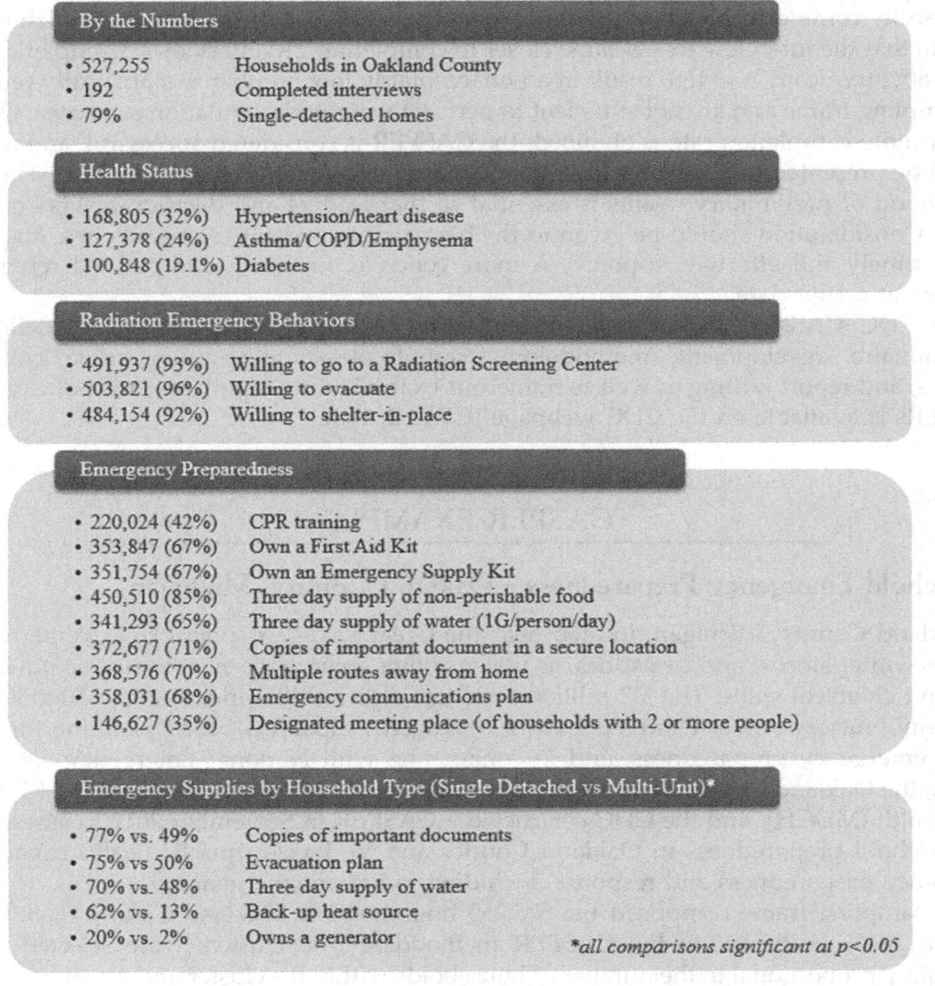

By the Numbers
• 527,255 Households in Oakland County
• 192 Completed interviews
• 79% Single-detached homes
Health Status
• 168,805 (32%) Hypertension/heart disease
• 127,378 (24%) Asthma/COPD/Emphysema
• 100,848 (19.1%) Diabetes
Radiation Emergency Behaviors
• 491,937 (93%) Willing to go to a Radiation Screening Center
• 503,821 (96%) Willing to evacuate
• 484,154 (92%) Willing to shelter-in-place
Emergency Preparedness
• 220,024 (42%) CPR training
• 353,847 (67%) Own a First Aid Kit
• 351,754 (67%) Own an Emergency Supply Kit
• 450,510 (85%) Three day supply of non-perishable food
• 341,293 (65%) Three day supply of water (1G/person/day)
• 372,677 (71%) Copies of important document in a secure location
• 368,576 (70%) Multiple routes away from home
• 358,031 (68%) Emergency communications plan
• 146,627 (35%) Designated meeting place (of households with 2 or more people)
Emergency Supplies by Household Type (Single Detached vs Multi-Unit)*
• 77% vs. 49% Copies of important documents
• 75% vs 50% Evacuation plan
• 70% vs. 48% Three day supply of water
• 62% vs. 13% Back-up heat source
• 20% vs. 2% Owns a generator
*all comparisons significant at p<0.05

FIGURE 7.3 Community Assessment for Public Health Emergency Response results, Oakland County, Michigan—2012 (CDC, 2013).

majority of households had basic emergency supplies of food, water, first aid, and emergency kits. However, when comparing single-detached homes versus multiunit dwellings, households in multiunit dwellings were less likely to have certain recommended emergency plans and supplies compared with those in single-detached homes (Murti et al., 2014).

Identifying and understanding the barriers that households may face in trying to improve their level of emergency preparedness can influence emergency planning and public health. The CASPER provided evidence that housing type is associated with having emergency plans and owning certain emergency supplies. This information allows public health officials to improve household emergency preparedness by targeting interventions to those owning and living in multiunit dwellings. OCHD and MCDH used results from the CASPER to update their state and county emergency and communication plans.

CONCLUSION

The ability of CASPER to quickly provide statistically valid data to leaders and decision-makers can have an important impact throughout the disaster life cycle as well as in nondisaster situations. The type and number of scenarios where CASPER is useful continues to expand throughout the United States (Bayleyegn et al., 2015). For example, CASPER data have been used to initiate public health action, facilitate disaster planning, and assess new or changing needs during the disaster recovery period. It has been also used to assess public health perceptions, estimate the needs of a community, assist in planning for emergency response, and as part of the public health accreditation process. Recommendations based on CASPER data result in public health actions that benefit the community.

References

Bayleyegn, T., Schnall, A., Ballou, S., Zane, D., Burrer, S., Noe, R., et al. (2015). Use of Community Assessments for Public Health Emergency Responses (CASPERs) to rapidly assess public health issues – United States, 2013–2012. *Prehospital and Disaster Medicine, 30*, 1–8.

Bayleyegn, T., Vagi, S., Schnall, A., Podgornik, M., Noe, R., & Wolkin, A. (2012). *Community Assessment for Public Health Emergency Response (CASPER) toolkit* (2nd ed).

Centers for Disease Control and Prevention (CDC). (2009). *Public health emergency response to the major ice storm in Kentucky—2009* (Unpublished final report). Atlanta, GA.

Centers for Disease Control and Prevention (CDC). (2013). *Household emergency preparedness, Oakland county, Michigan, September 2012*. Retrieved from http://www.michigan.gov/documents/mdch/MI_CASPER_Report_FINAL_02112013_413612_7.pdf.

Centers for Disease Control and Prevention (CDC). (2016a). *Interactive map of CASPERs*. Retrieved from https://www.cdc.gov/nceh/hsb/disaster/casper/casper_map.htm.

Centers for Disease Control, Prevention (CDC). (2016b). *Sampling methodology: Geographic Information System (GIS) CASPER toolbox*. Retrieved from https://www.cdc.gov/nceh/hsb/disaster/casper/sampling.htm.

Federal Emergency Management Agency (FEMA). (2001). *Rapid needs assessment in federal disaster operations: Operation manual*. Retrieved from https://www.hsdl.org/?view&did=4199.

Horney, J. A., MacDonald, P. D. M., Van Willigen, M., Berke, P. R., & Kaufman, J. S. (2010a). Individual actual or perceived property flood risk: Did it predict evacuation from Hurricane Isabel in North Carolina, 2013? *Risk Analysis, 30*, 501–511.

Horney, J. A., MacDonald, P. D. M., Van Willigen, M., Berke, P. R., & Kaufman, J. S. (2010b). Factors associated with evacuation from Hurricane Isabel in North Carolina, 2003. *International Journal of Mass Emergencies and Disasters, 28*(1), 33–58.

International Federation of Red Cross and Red Crescent Societies (IFRC). (2008). *Guidelines for assessment in emergencies*. Retrieved from http://www.ifrc.org/Global/Publications/disasters/guidelines/guidelines-for-emergency-en.pdf.

International Federation of Red Cross and Red Crescent Societies (IFRC). (n.d.). Emergency needs assessment. Retrieved from http://www.ifrc.org/en/what-we-do/disaster-management/responding/disaster-response-system/emergency-needs-assessment/.

Lemeshow, S., & Robinson, D. (1985). Surveys to measure programme coverage and impact: A review of the methodology used by the expanded programme on immunization. *World Health Statistics, 1*, 65–75.

Malilay, J. M., Flanders, W. D., & Brogan, D. (1996). A modified cluster-sampling method for post-disaster rapid assessment of needs. *Bulletin of the World Health Organization, 74*, 399–405.

Malilay, J. M., Heumann, M., Perrotta, D., Wolkin, A., Schnall, A., Podgornik, M. N., et al. (2014). The role of applied epidemiology methods in the disaster management cycle. *American Journal of Public Health, 104*, 2092–2102.

Murti, M., Bayleyegn, T., Stanbury, M., Blies, S., Flanders, D., Yard, E., et al. (2014). Household emergency preparedness of housing type from a Community Assessment for Public Health Emergency Response (CASPER), Michigan. *Disaster Medicine and Public Health Preparedness, 8*, 12–19.

Pan American Health Organization. (2014). *Rapid needs assessment*. Retrieved from http://www.paho.org/disasters/index.php?option=com_content&view=article&id=744%3Arapid-needs-assessment&Itemid=800&lang=en.

United States Agency for International Development. (2014). *A rapid needs assessment guide: For education in countries affected by crisis and conflict*. Retrieved from http://eccnetwork.net/wp-content/uploads/USAID-RNAG-FINAL.pdf.

U.S. Census Bureau. (2010). QuickFacts. Retrieved from http://www.census.gov/quickfacts/table/RHI105210/26125.

Wallace, J. W., Poole, C., & Horney, J. A. (2016). The association between actual and perceived flood risk and evacuation from Hurricane Irene in Beaufort County, North Carolina. *Journal of Flood Risk Management, 9*, 125–135.

World Health Organization (WHO). (2005). *Immunization coverage cluster survey – reference manual*. Retrieved from http://apps.who.int/iris/bitstream/10665/69087/1/WHO_IVB_04.23.pdf.

CHAPTER

8

Applications: Assessment of Chemical Exposures

Epidemiologic Investigations After Large-Scale Chemical Releases

Maureen F. Orr, Mary Anne Duncan[†]

Agency for Toxic Substances and Disease Registry, Atlanta, GA, United States

BACKGROUND

The Agency for Toxic Substances and Disease Registry (ATSDR), based in Atlanta, Georgia, is a federal public health agency of the U.S. Department of Health and Human Services. Both ATSDR and the Centers for Disease Control and Prevention's (CDC) National Center for Environmental Health (NCEH) operate under the same office of the director. Safeguarding communities from chemical exposures is ATSDR's ultimate goal, and agency staff work closely with other federal, state, and local agencies; tribal governments; local communities; and health-care providers to fulfill that goal (ATSDR, 2015a). For over 20 years, ATSDR has partnered with state health departments to perform surveillance for acute chemical release incidents. In 2010, as part of its chemical incident surveillance program, the National Toxic Substance Incidents Program, ATSDR added the Assessment of Chemical Exposures (ACE) program component. ATSDR designed the ACE program to provide expert epidemiologic and toxicological assistance and other resources to state and local health departments for performing rapid assessments during the aftermath of large-scale chemical releases resulting in multiple casualties. Resources include standardized survey forms and data entry programs, geospatial analysis tools, and laboratory specimen testing. ATSDR built the ACE program based on lessons learned from the Graniteville, South Carolina, train derailment that released chlorine gas (Duncan & Orr, 2010). Community exposure to the release was widespread and resulted in significant acute health effects and substantial health-care needs. In the years since the incident, researchers have conducted multiple investigations

[†]Deceased

Disaster Epidemiology
http://dx.doi.org/10.1016/B978-0-12-809318-4.00010-1

to understand the impact of this incident and determine actions that could have minimized it (Dunning & Oswalt, 2007; Svendsen et al., 2012).

Lacking sufficient personnel to complete a timely investigation immediately after the Graniteville event, the South Carolina Department of Health and Environmental Control (SC DHEC) requested assistance from CDC/ATSDR to perform an epidemiologic assessment of the chlorine exposure and resulting health effects (Wenck et al., 2007). As part of the assessment, they performed a health survey. Had the ACE program been in existence, the SC DHEC team would have had immediate access to a wealth of assessment resources, including a standardized survey and data entry program. Also, federal staff could have supported the investigation from the beginning.

Five months after the incident, an interagency team resurveyed participants to determine their health status and needs to assist in planning additional interventions in the community. Many participants reported that they continued to experience physical symptoms and require medical care. Some also reported psychological symptoms indicative of posttraumatic stress disorder. These results led to the survey team's recommendation that investigators collect self-reported symptoms as soon as possible after a chemical exposure, because accurate recall might diminish within months (Duncan et al., 2011).

A survey of health screening participants 8 to 10 months after the acute chlorine exposure revealed that participants underreported respiratory symptoms when compared with simultaneous abnormal spirometry results. Because agreement between perceived and measured respiratory symptoms was low, the researchers determined that relying on the self-reported questionnaire was not adequate to objectively assess the lung health of the population following irritant gas exposure (Clark et al., 2013). For this reason, ACE often employs different measures of the severity of health effects and strives to interview injured persons within 2 weeks of their exposures. In addition, ACE tries to include objective measures, such as spirometry, medical records, and biological testing, if available, in subsequent investigations.

ASSESSMENT OF CHEMICAL EXPOSURES RESOURCES

The ACE Toolkit (NTSIP, 2016) contains surveys that are readily customizable to each situation, including individual surveys with sections for adults, including responders, and children; household surveys; a hospital survey; as well as ATSDR Rapid Response Registry materials, a consent form, and a medical chart abstraction form. The individual survey also includes a section for collecting data on household pets and a veterinary chart abstraction form. Many of the surveys and consent forms have been translated into Spanish. Epi Info databases with an ACE data management guide are available to assist with data entry and management. Training materials are also in the toolkit, including an interviewer training manual and the ACE workbook used during the ACE training courses.

The ACE team is available to provide technical assistance by phone or email with assessments of chemical incidents. If on site assistance is needed, the ACE program uses the Epi-Aid mechanism to send personnel to assist state health departments in an epidemiologic response. On receiving an invitation from the state epidemiologist, ATSDR deploys Epidemic Intelligence Service officers and other staff to support the state's needs. Teams can include specialized personnel such as an industrial hygienist or medical toxicologist to supplement the skill sets of staff in the inviting agency. Generally, ACE teams consult with geographers for geospatial analysis needs and also with environmental public health laboratorians for guidance on biologic sampling. The team can be deployed within 2 days of receiving a request for assistance, at no charge to the inviting agency (ATSDR, 2015b).

ATSDR may also assist inviting agencies in developing a public report to inform the community, present at public health meetings, or publish in the peer-reviewed scientific literature to share lessons learned during the assessment. The CDC Morbidity and Mortality Weekly Report (MMWR), which has published the findings of several ACE investigations, is a timely and high-profile way to distribute ACE findings to medical and public health professionals, educators, and researchers nationwide.

The ACE program seeks to build capacity among public health professionals by conducting in-person trainings, as well as offering free training materials online. An ACE training course that offers free continuing education credit for several professions for performing investigations after chemical spills is available through CDC/ATSDR. ACE trainings are sometimes offered in conjunction with other disaster epidemiology trainings, including the NCEH Community Assessment for Public Health Emergency Response (CASPER) training and the National Institute for Occupational Safety and Health (NIOSH) Emergency Responder Health Monitoring and Surveillance training; these trainings are generally available on request (Duncan, 2014) (Fig. 8.1).

State health departments have used ACE resources in a variety of chemical incidents. Since each situation was different, the investigation activities varied. Approaches have included interviewing responders; interviewing attending hospital staff; administering occupational,

FIGURE 8.1 Conducting household surveys after vinyl chloride incident in New Jersey. In photo: *Left*: Jason Wilken (Epidemic Intelligence Service (EIS) officer); *Right*: Kim Brinker (EIS officer). *Photo by Mary Anne Duncan.*

individual, and household surveys; and conducting medical chart abstractions. The ACE program, tools, and materials made these activities relatively quick and easy to initiate. In these investigations, ACE staff partnered with other federal agencies, including the CDC NIOSH, NCEH, the National Institute of Environmental Health Sciences (NIEHS) at the National Institute of Health, and the US Environmental Protection Agency (EPA). Because many of these incidents were high profile, ATSDR staff supported local and state health departments with crisis communication, liaisons with other federal agencies, dissemination of preliminary findings, and suggested action plans during the response.

OUTCOMES OF ASSESSMENT OF CHEMICAL EXPOSURES INVESTIGATIONS

The assessment of chemical incidents has many potential benefits for community members, state and/or local health departments, emergency responders, and federal agencies. The main benefits from implementing ACE assessments include identifying the following:

1. Actions to prevent future incidents
2. An exposed cohort of individuals for further follow-up
3. Areas for improvement in public health and safety response
4. Effects on the community and target audiences for information and assistance
5. Lessons that other health departments, emergency departments, researchers, first responders, and hospitals can learn from to improve future responses

The following examples demonstrate these benefits:

- In attempting to recycle a closed tank, a metal recycling facility caused a chlorine release that affected employees, customers, and nearby residents. An ACE team assisted the state health department in conducting an epidemiologic assessment of the release. The main outcome of the investigation was a chemical release alert warning of the dangers posed by closed containers sold for recycling. The warning included instructions to follow if members of the public encountered a container and directions for evacuating upwind in the event of a chemical release. This alert, which was also translated into Spanish, was sent to all the recycling facilities in the state and distributed through a recycling industry organization. MMWR also published an article about the release and resulting alert for the public health audience (Kelsey et al., 2011). The state health department later followed up with individuals affected by the release; staff found that some were still experiencing ill effects and routed them to appropriate health care.
- At a Caribbean resort, a vacationing family fell gravely ill with the same symptoms. They were transported to the hospital by first responders. Fortuitously, clinicians suspected the source of the problem—an acute chemical exposure. When the resort was questioned, it was found that there had been a recent pesticide application. When EPA conducted an investigation, it was discovered that a very dangerous pesticide methyl bromide had been used illegally indoors. The U.S. Virgin Island's Health Department requested ACE assistance with surveying potentially affected workers, vacationers, and

responders to assess any health effects. This ACE assessment served as a reminder to clinicians to consider the possibility of acute chemical toxicity in relevant clinical and epidemiologic situations; prompt identification of an exposure can prevent further exposure and subsequent illness. This assessment also recommended implementing safer pesticide handling practices—in this case not using outdoor pesticides indoors—and warning HazMat responders about potential exposure in a timely manner so they can seek medical evaluation. MMWR published the findings of this assessment to alert the public health and medical community (Kulkarni et al., 2015).

- Another ACE assessment surveyed employees who were working downwind from a refrigeration facility at a Deepwater Horizon Gulf oil spill cleanup site and were exposed to ammonia. The ACE team conducted a survey on the health effects experienced as a result of the ammonia exposure and arranged for NIEHS to include the exposed employees in the Gulf Long-Term Follow-Up Study. In addition, the ACE program participated in an after-action review with the responders to this incident. The review revealed that the alarm notification system had failed, so people in the area were unaware of the release, possibly explaining why the cleanup workers did not evacuate or shelter-in-place. The county later installed a reverse 9-1-1 system to call residential and business telephones in a defined geographic area and deliver recorded emergency notification (ATSDR, 2015b).
- In another investigation, ACE assessed the effects of an accidental chemical mixing incident that caused a chlorine gas release at a poultry processing facility. The team interviewed the responders and the hospital and noted that hospitals expressed concerns about the lack of communication from the scene concerning patient condition and arrival. The investigation revealed language barriers and triaging shortfalls that caused confusion when patients' relatives could not locate them. Health department staff were prepared to communicate with responding hospital emergency departments to facilitate patient triage and care and provide translating services to the large number of non–English-speaking patients. However, because responders did not immediately notify the health department of the incident, these resources were not available to assist in the response. The ACE team learned that timely notification did not occur because the emergency plan threshold number of casualties for alerting the health department was too high. This experience led the state health department, working with the state department of emergency management, to modify the notification procedures by lowering the notification level (Christensen et al., 2016). Two weeks after the new procedures went into effect, two ammonia releases occurred on the same day. Thanks to the modified procedures, emergency responders notified the health department immediately, allowing the department to assist in a well-run response. ATSDR notified NIOSH of the incident, and NIOSH decided to conduct a concurrent Health Hazard Evaluation of the exposed workers in the poultry processing facility. The evaluation prompted an MMWR warning about safe labeling and handling of chemicals (Whitlow et al., 2012). NIOSH later conducted a follow-up health assessment and pulmonary function tests on workers to evaluate longer-lasting effects of the chlorine exposure.

- ACE investigated a chemical incident in which a tank containing chemicals used in coal processing leaked into a river just upstream from the intake of the municipal water supply for a city of approximately 300,000 people. The health department issued a "Do Not Use" water order for a nine-county area and requested an ACE assessment. The ACE team surveyed area hospitals to learn about their experiences with the "Do Not Use" water order and abstracted and analyzed hospital charts of potentially exposed persons. In addition, the ACE team reviewed the existing disaster epidemiology capacity within the inviting agency. The hospital survey findings provided information to hospitals planning for emergencies that may compromise their water supply. The investigation revealed that the amount of water needed to sterilize equipment for surgery and to provide clean linens and food services was higher than originally expected. The health department used the existing epidemiology capacity report in planning future disaster responses and in disaster epidemiology training. Results from the hospital chart reviews were used in conjunction with results of an NCEH CASPER for local outreach and education efforts to alleviate public concerns about spill-related health effects and using public water (CDC, 2014).
- ACE was involved in another water-related incident during the investigation of lead levels in Flint, Michigan drinking water. Some residents reported rashes (or worsening rashes) to their health-care providers and believed that their water supply was the cause. In response, ACE assisted the state health department in an investigation that included three steps: (1) a survey of affected individuals, (2) a dermatologist's evaluation of their rashes, and (3) water quality tests performed in their residences. Investigators found no pattern of water contaminants or water quality parameters in the homes sampled for this investigation that could account for the rashes. However, the investigation did reveal previously reported water issues that may be linked to the problems. The findings provided some assurance to residents that they could safely use their water again for bathing, since many of them had limited or stopped bathing during the crisis (Unified Coordination Group-Flint, Michigan, 2016).
- During a train derailment in New Jersey, a punctured tanker car (Fig. 8.2) released approximately 24,000 gallons of vinyl chloride at the edge of a small town. The

FIGURE 8.2 Hole in tanker car from vinyl chloride incident in New Jersey. *Photo by: Mary Anne Duncan.*

incident commander issued a community shelter-in-place order and expanded it to surrounding areas. However, the order was lifted and reestablished repeatedly over the next 4 days, as vinyl chloride levels in the air fluctuated due to weather conditions. To answer community concerns about possible exposure to vinyl chloride, the ACE team, in partnership with the state and local health department, surveyed community members who were potentially exposed and performed hospital chart abstractions. State partners also mailed a survey to all households in the community. To address community concerns, the New Jersey Department of Health (NJDH) published a final report stating that residents closest to the spill were the most affected (New Jersey Department of Health, 2014). Because of uncertainty about the long-term effects of this level of exposure, the health department advised residents to seek routine, age-appropriate health care.

Additionally, to reduce the public health impact of future chemical exposure events NJDH recommended the following: (1) that employers of emergency responders and incident commanders follow established regulations and guidance regarding the use of personal protective equipment, (2) that schools design emergency response plans to protect children and staff from chemical exposures, and (3) that emergency responders evacuate residents close to an exposure scene as soon as feasible.

The health surveys found that Paulsboro, New Jersey, residents received information about the derailment and instructions about safety precautions from a variety of sources, and many wanted more direct and consistent communication from a local authority. The ACE team recommended that local officials prepare and make available community-specific emergency planning educational materials tailored to relevant catastrophic hazards. This event is another instance when emergency response did not notify and include the state health department initially; therefore, the team recommended that public health agencies be engaged within the incident command system to provide guidance and address health concerns of the community.

CONCLUSION

The ACE program has repeatedly demonstrated its value in assisting state and local health departments with epidemiologic assessments after chemical incidents. In addition, by responding quickly, using structured tools and methods, and distributing findings to the medical, public health, training, research, and public safety community in a timely manner, ACE advances the current body of knowledge of chemical incident preparedness, response, recovery, and prevention.

Disclaimer

The findings and conclusions in this report are those of the author(s) and do not necessarily represent the views of the Agency for Toxic Substances and Disease Registry.

References

ATSDR. (2015a). *Safeguarding communities from chemical exposures*. Retrieved from http://www.atsdr.cdc.gov/docs/APHA-ATSDR_book.pdf.

ATSDR. (2015b). *NTSIP, incident investigations*. ACE. Retrieved from http://www.atsdr.cdc.gov/ntsip/ace.html.

Centers for Disease Control and Prevention (CDC). (2014). *Disaster response and recovery needs of communities affected by the Elk River chemical spill, West Virginia*. Retrieved from http://www.dhhr.wv.gov/News/2014/Documents/WVCASPERReport.pdf.

Christensen, B., Duncan, M. A., King, S. C., Hunter, C., Ruckart, P., & Orr, M. F. (2016). Challenges during a chlorine gas emergency response. *Disaster Medicine and Public Health Preparedness, 10*, 553–555.

Clark, K. A., Chanda, D., Balte, P., Karmaus, W. J., Cai, B., Vena, J., et al. (2013). Respiratory symptoms and lung function 8-10 months after community exposure to chlorine gas: A public health intervention and cross-sectional analysis. *BMC Public Health, 13*, 945–956.

Duncan, M. A. (2014). Assessment of chemical exposures: Epidemiologic investigations after large-scale chemical releases. *Journal of Environmental Health, 77*, 36–38.

Duncan, M. A., Drociuk, D., Belflower-Thomas, A., Van Sickle, D., Gibson, J. J., Youngblood, C., et al. (2011). Follow-up assessment of health consequences after a chlorine release from a train derailment—Graniteville, SC, 2005. *Journal of Medical Toxicology, 7*, 85–91.

Duncan, M. A., & Orr, M. F. (2010). Evolving with the times, the new national toxic substance incidents program. *Journal of Medical Toxicology, 6*, 461–463.

Dunning, A. E., & Oswalt, J. L. (2007). Train wreck and chlorine spill in Graniteville, South Carolina: Transportation effects and lessons in small-town capacity for no-notice evacuation. *Transportation Research Record: Journal of the Transportation Research Board, 2009*, 130–135.

Kelsey, K., Roisman, R., Kreutzer, R., Kelsey, K., Roisman, R., Kreutzer, R., et al. (2011). Chlorine gas exposure at a metal recycling facility — California, 2010. *Morbidity and Mortality Weekly Report, 60*, 951–954.

Kulkarni, P. A., Duncan, M. A., Watters, M. T., Graziano, L. T., Vaouli, E., Cseh, L. F., et al. (2015). Severe illness from methyl bromide exposure at a condominium resort — U.S. Virgin Islands, March 2015. *Morbidity and Mortality Weekly Report, 64*, 763–766.

National Toxic Substance Incidents Program (NTSIP). ACE toolkit. (February 2, 2016). Retrieved from http://www.atsdr.cdc.gov/ntsip/ace_toolkit.html.

New Jersey Department of Health. (2014). *Paulsboro train derailment and vinyl chloride release, November 30, 2012 health survey findings and air quality impacts*. Retrieved from http://www.nj.gov/health/ceohs/documents/pau/factsheet.pdf.

Svendsen, E. R., Runkle, J. R., Ramana Dhara, V., Lin, S., Naboka, M., Mousseau, T. A., et al. (2012). Epidemiologic methods lessons learned from environmental health disasters: Chernobyl, the World Trade Center, Bhopal, and Graniteville, South Carolina. *International Journal of Environmental Research and Public Health, 9*, 2894–2909.

Unified Command Group- Flint Michigan. (2016). *Flint rash investigation*. Retrieved from http://www.phe.gov/emergency/events/Flint/Documents/rash-report.pdf.

Wenck, M. A., Van Sickle, D., Drociuk, D., Belflower, A., Youngblood, C., Whisnant, M. D., et al. (2007). Rapid assessment of exposure to chlorine released from a train derailment and resulting health impact. *Public Health Reports, 122*, 784–792.

Whitlow, A., Louie, S., Mueller, C., King, B., Page, E., Bernard, B., et al. (2012). Chlorine gas release associated with employee language barrier — Arkansas, 2011. *Morbidity and Mortality Weekly Report, 61*, 981–985.

Memoriam

This chapter is dedicated to the contributions of Commander Mary Anne Duncan, DVM, MPH, DACVPM 1963–2017. Mary Anne was instrumental in getting the ACE program up and running and making it an instrumental program at ATSDR for environmental public health disasters.

Vignette: Geothermal Venting and Emergency Preparedness—Lake County, California

Svetlana Smorodinsky[1], *Tracy Barreau*[1], *Jason Wilken*[2,3]
[1]California Department of Public Health, Richmond, CA, United States; [2]Centers for Disease Control and Prevention, Atlanta, GA, United States; [3]U.S. Public Health Service Commissioned Corps, Rockville, MD, United States

Lake County, California, population 64,665 (US Census Bureau, 2010), is located about 100 miles north of San Francisco, in a region prone to natural disasters (floods, wildfires) (Federal Emergency Management Agency) and notable for volcanic, seismic, and geothermal activity (USGS Volcano Hazards Program). The same geologic and geographic characteristics that make geothermal energy a viable industry in Lake County—the largest complex of geothermal power plants in the world is located there (California Energy Commission, 2016)—also makes Lake County vulnerable to a variety of environmental hazards such as earthquakes, volcanic eruptions, and venting of harmful gases, including hydrogen sulfide (H_2S), methane (CH_4), carbon dioxide, and radon (County of Lake, 2013).

Venting of geothermal gases has been documented in the City of Clearlake, the largest city in Lake County, and at "Gas Hill" in Kelseyville (Lake County Community Development Department). Instances of spontaneous localized H_2S and CH_4 releases have resulted in household evacuations, home demolition, a health advisory, and implementation of engineering efforts to mitigate the worst areas of known gas releases (Lake County News Reports, 2010; Lake County Air Quality Management District, 2011; Lake County Office of Emergency Services, 2012).

The extent of geothermal gas venting in the rest of the county has been largely unexplored (Lake County Community Development Department; Lake County Office of Emergency Services, 2012). Signs of geothermal venting include bubbling in water puddles, unusual rusting patterns on fence posts and other metal objects, and rotten egg odors (CDPH, 2012). County residents have periodically observed these signs throughout the county (Staff Reports, 2011).

Disaster Epidemiology
http://dx.doi.org/10.1016/B978-0-12-809318-4.00011-3

Additionally, tectonic activity can cause new gas vents to form or change the location or intensity of existing vents. In fact, Clearlake experienced >1200 earthquakes magnitude 1.5 or greater in 2015) (Earthquake Track).

Because of these potential environmental threats, Lake County officials wanted to understand the degree of the community's experiences with geothermal venting and the level of residents' disaster preparedness. In November 2012, Lake County Public Health Division, California Department of Public Health, and the Centers for Disease Control and Prevention partnered to conduct a CASPER in Lake County. The goals of the CASPER were to assess experience and perceptions associated with geothermal gas venting among county residents, to conduct air monitoring to identify areas of potential concern for geothermal venting and vapor intrusion, and to assess residents' disaster preparedness.

The CASPER sampling frame contained 26,730 housing units located in 12 cities and towns in Lake County, which were chosen to represent the most populated areas of the county, and not the entire county as a whole. The CASPER survey included questions on household demographics, disaster preparedness, private wells, and experiences with and perceptions of geothermal gas venting. The interview teams also recorded observational data on evidence of geothermal venting outside the sampled homes (such as unexpected rust patterns on fence posts, a rotten egg smell, or bubbling in puddles). Teams provided all approached households with educational materials regarding emergency preparedness and other public health topics of local relevance (such as well water information, mercury in fish, and radon safety).

The Lake County CASPER was the first in the nation to conduct environmental sampling as part of the assessment. An air measurement component was incorporated, reflecting the recognition that geothermal gases were intruding into structures in the City of Clearlake and the need to identify other locations in the county where the gases may be diffusing from the subsurface. Air monitoring teams conducted air sampling for H_2S and CH_4 in the same CASPER clusters, taking spot measurements in and near water meter vaults (enclosed spaces where gases can accumulate) and outside of systematically selected homes (Chiu et al., 2015).

CASPER results are presented in Fig. 1 (CDPH, 2012).

All H_2S measurements in water meter vaults were $\leq$1 parts per billion (ppb). Aboveground median values per CASPER cluster were 0–4 ppb (in comparison, H_2S concentrations in urban areas are generally <1 ppb) (Agency for Toxic Substances and Disease Registry (ATSDR), 2006). All CH_4 measurements were 0% of the lower explosive limit (% LEL), with the exception of two readings measured at 1% LEL. The highest measured H_2S and CH_4 levels were detected in several City of Clearlake clusters, suggesting the venting may be sporadic or highly localized (Chiu et al., 2015).

This CASPER was a successful collaboration of local, state, and federal public health agencies and the first in the nation to incorporate air monitoring. Data generated by this assessment were useful in revision of county emergency plans and improvements in communications. For example, most residents reported having pets or livestock, and most reported that they would take them along during a mandatory evacuation, which prompted the county to address the need for animal-friendly shelters. During the devastating Valley Fire in 2015 (CalFire, 2015), the county Animal Care and Control Department

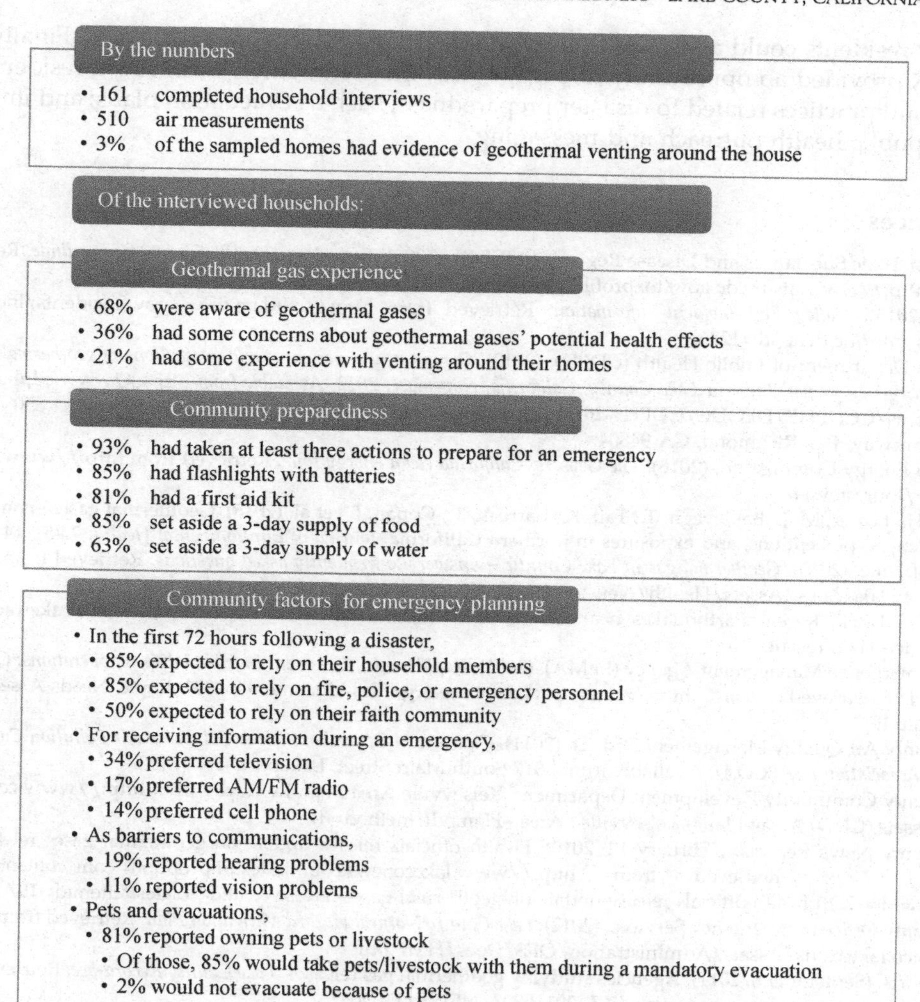

By the numbers
• 161 completed household interviews
• 510 air measurements
• 3% of the sampled homes had evidence of geothermal venting around the house
Of the interviewed households:
Geothermal gas experience
• 68% were aware of geothermal gases
• 36% had some concerns about geothermal gases' potential health effects
• 21% had some experience with venting around their homes
Community preparedness
• 93% had taken at least three actions to prepare for an emergency
• 85% had flashlights with batteries
• 81% had a first aid kit
• 85% set aside a 3-day supply of food
• 63% set aside a 3-day supply of water
Community factors for emergency planning
• In the first 72 hours following a disaster,
• 85% expected to rely on their household members
• 85% expected to rely on fire, police, or emergency personnel
• 50% expected to rely on their faith community
• For receiving information during an emergency,
• 34% preferred television
• 17% preferred AM/FM radio
• 14% preferred cell phone
• As barriers to communications,
• 19% reported hearing problems
• 11% reported vision problems
• Pets and evacuations,
• 81% reported owning pets or livestock
• Of those, 85% would take pets/livestock with them during a mandatory evacuation
• 2% would not evacuate because of pets

FIGURE 1 CASPER results, Lake County, California–2012.

and its animal rescue volunteers provided animal support adjacent to shelters and veterinary disaster healthcare volunteers assisted local veterinarians with injured animal treatment.

Lake County Office of Emergency Services distributed hundreds of public alert radios, some of which were designed for people with sensory impairments, to at-risk residents. Poor cell coverage and other communication challenges during CASPER implementation led to improvements in ham radio operations and eventual installation of a cell tower in one of the remote communities. The air monitoring identified no immediate risk to the sampled communities; however, continued vigilance and reporting of characteristic H_2S

signs by residents could assist the County in identifying new geothermal vents. Finally, this CASPER provided an opportunity to engage with the community; learn about residents' attitudes and practices related to disaster preparedness, such as evacuation plans; and improve county public health outreach and messaging.

References

Agency for Toxic Substances and Disease Registry (ATSDR). (2006). *Toxicological profile for hydrogen sulfide*. Retrieved from http://www.atsdr.cdc.gov/toxprofiles/tp114.pdf.

CalFire. (2015). *Valley Fire incident information*. Retrieved from http://cdfdata.fire.ca.gov/incidents/incidents_details_info?incident_id=1226.

California Department of Public Health (CDPH). (2012). *Community experiences and perceptions of geothermal venting and emergency preparedness in Lake County, California, November, 2012. Available from* https://www.cdph.ca.gov/Programs/CCDPHP/DEODC/CDPH%20Document%20Library/Lake%20CASPER%20report.pdf, 850 Marina Bay Parkway, P-3, Richmond, CA 94804.

California Energy Commission. (2016). *The Geysers: California clean energy tour*. Retrieved from http://www.energy.ca.gov/tour/geysers.

Chiu, C. H., Lozier, M. J., Bayleyegn, T., Tait, K., Barreau, T., Copan, L., et al. (2015). Geothermal gases—community experiences, perceptions, and exposures in northern California. *Journal of Environmental Health, 78*(5), 14—21.

County of Lake. (2013). *Geothermal gas in Lake County – answers to frequently asked questions*. Retrieved from http://health.co.lake.ca.us/Assets/Health/New+geothermal+gas+informational+brochure.pdf.

Earthquake Track. Recent Earthquakes near Clearlake, California. Retrieved from http://earthquaketrack.com/us-ca-clearlake/recent.

Federal Emergency Management Agency (FEMA). (2001). *Rapid needs assessment in federal disaster operations: Operation manual*. Retrieved from http://www.parkdatabase.org/files/documents/2001_Rapid-Needs-Assessment_FEMA.pdf.

Lake County Air Quality Management District. (2011). *Clearlake H_2S, presentation to California Air Pollution Control Officers Association (CAPCOA)*. Available from 2617 South Main Street, Lakeport, CA 95453.

Lake County Community Development Department. Kelseyville Area Plan. Retrieved from http://www.co.lake.ca.us/Assets/CDD/Area+Plans/Kelseyville+Area+Plan.pdf?method=1.

Lake County News Reports. (February 12, 2010). Health officials release update on geothermal gases release. *Lake County News*. Retrieved from http://www.lakeconews.com/index.php?option=com_content&view=article&id=12581:health-officials-release-update-on-geothermal-gases-release-&catid=1:latest&Itemid=197.

Lake County Office of Emergency Services. (2012). *Lake County Natural Hazard Mitigation Plan*. Retrieved from http://www.co.lake.ca.us/Assets/Administration/OES/Docs/HMP.pdf.

Staff Reports. (September 8, 2011). Agencies studying geothermal gas releases. *Lake County Record-bee*. Retrieved from http://www.record-bee.com/article/ZZ/20110908/NEWS/110909237.

US Census Bureau. (2010). *QuickFacts*. Retrieved from https://www.census.gov/quickfacts/table/PST045215/06033,00.

USGS Volcano Hazards Program. Clearlake Volcanic Field. Retrieved from https://volcanoes.usgs.gov/volcanoes/clear_lake/.

Vignette: CASPER in Response to a Slow-Moving Disaster—The California Drought

Svetlana Smorodinsky[1], *Jason Wilken*[1], *Tracy Barreau*[2,3]

[1]California Department of Public Health, Richmond, CA, United States; [2]Centers for Disease Control and Prevention, Atlanta, GA, United States; [3]U.S. Public Health Service Commissioned Corps, Rockville, MD, United States

California's unprecedented drought, which began in 2012, reduced available surface water and depleted groundwater aquifers. Thousands of private wells have gone dry and millions of trees have died in drought-stricken forests. Governor Edmund G. Brown Jr. proclaimed a State of Emergency in January 2014 (Office of the Governor, 2014) and, as of November 2015, California recorded 63 Emergency Proclamations from city, county, and tribal governments, and special districts (Office of the Governor, 2015).

Tulare County, a largely agricultural county in the heart of California's Central Valley, has had the majority of reported private well failures in the state (Governor's Office of Emergency Services, 2015). Tulare County implemented comprehensive water assistance programs including offering relocation assistance for residents with no running water in their homes, making available household potable and nonpotable water tanks, and offering emergency food assistance for residents impacted by the drought.

Mariposa County, located at the foothills of the Sierra Nevada Mountains, has experienced fewer private well failures than Tulare County, but the drought has severely impacted the County's forests, with an estimated 30%–50% mortality of pine, fir, and oak (Tafoya, 2016). Mariposa County made dry-well assistance available to residents impacted by the drought and promoted numerous household water conservation techniques, including graywater systems (i.e., any of several types of systems that reuse wastewater from showers, bathtubs, and clothes washing machines). Even though a County ordinance legalized graywater systems in 1991 (Mariposa County Ordinance 809, 1991), it was unclear whether residents were aware of these water conservation opportunities.

Disaster Epidemiology
http://dx.doi.org/10.1016/B978-0-12-809318-4.00012-5

The effects of drought on population health and behavioral health in developed nations have not been well characterized, and the extent of adoption of water conservation practices in the above counties was unknown. In October and November 2015, California Department of Public Health collaborated with Mariposa County Health Department and Tulare County Health and Human Services Agency to conduct CASPERs to better quantify drought-related population impacts and to inform public health decisions and actions. Tulare County conducted two simultaneous CASPERs, one in northern and one in southern parts of the county; the two sampling frames were designed to enrich for private well ownership. Mariposa chose a single county-wide sampling frame. The questionnaire included topics on household demographics; knowledge, attitudes, and practices regarding the drought; access to and use of water; water conservation practices; perceived impacts of the drought on health, mental health, and finances; and preferred communication methods.

CASPER results are presented in Fig. 1 (CDPH, 2015a, 2015b).

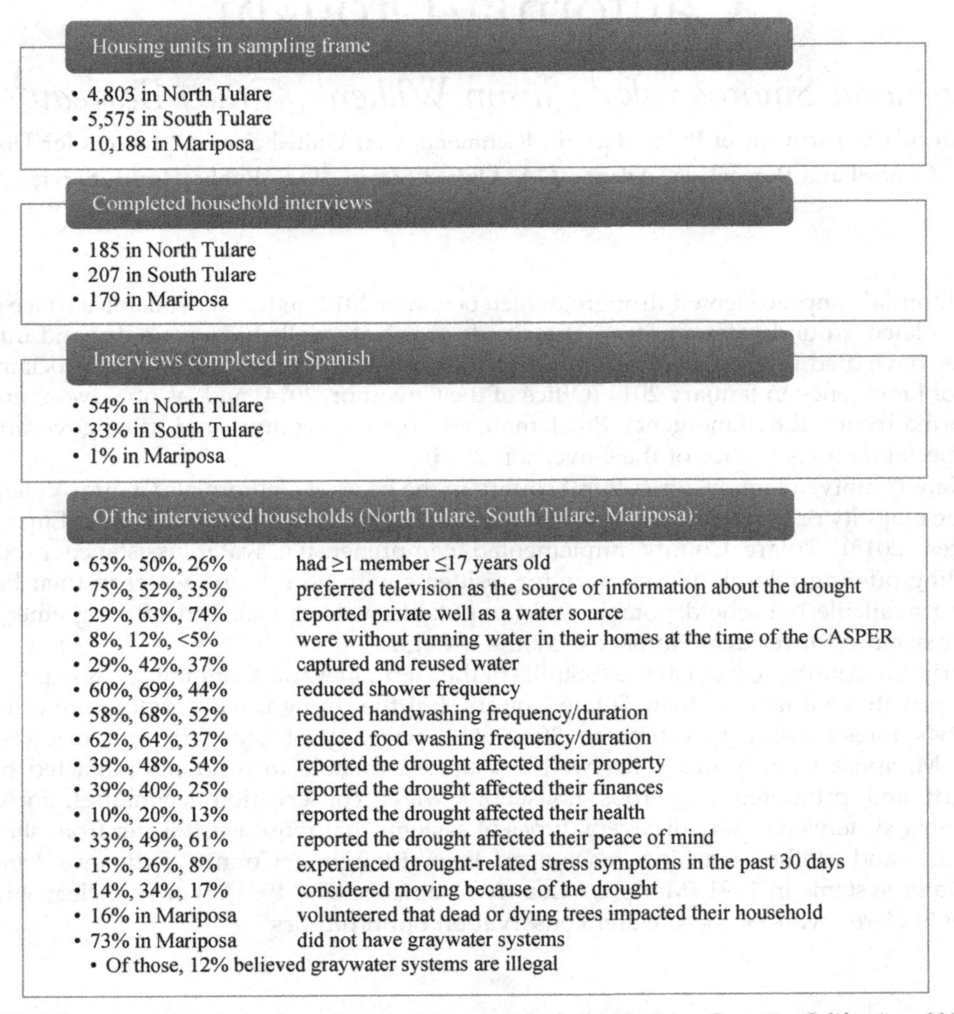

Housing units in sampling frame
• 4,803 in North Tulare
• 5,575 in South Tulare
• 10,188 in Mariposa
Completed household interviews
• 185 in North Tulare
• 207 in South Tulare
• 179 in Mariposa
Interviews completed in Spanish
• 54% in North Tulare
• 33% in South Tulare
• 1% in Mariposa
Of the interviewed households (North Tulare, South Tulare, Mariposa):
• 63%, 50%, 26% had ≥1 member ≤17 years old
• 75%, 52%, 35% preferred television as the source of information about the drought
• 29%, 63%, 74% reported private well as a water source
• 8%, 12%, <5% were without running water in their homes at the time of the CASPER
• 29%, 42%, 37% captured and reused water
• 60%, 69%, 44% reduced shower frequency
• 58%, 68%, 52% reduced handwashing frequency/duration
• 62%, 64%, 37% reduced food washing frequency/duration
• 39%, 48%, 54% reported the drought affected their property
• 39%, 40%, 25% reported the drought affected their finances
• 10%, 20%, 13% reported the drought affected their health
• 33%, 49%, 61% reported the drought affected their peace of mind
• 15%, 26%, 8% experienced drought-related stress symptoms in the past 30 days
• 14%, 34%, 17% considered moving because of the drought
• 16% in Mariposa volunteered that dead or dying trees impacted their household
• 73% in Mariposa did not have graywater systems
• Of those, 12% believed graywater systems are illegal

FIGURE 1 CASPER results, drought assessments in Tulare and Mariposa Counties, California – 2015.

A severe drought is a chronic disaster where outcomes are related to long exposures with far-reaching impacts on the economy, environment, and communities (including social cohesion) and leading to both direct and indirect health consequences (Stanke, Kerac, Prudhomme, Medlock, & Murray, 2013). Drought-impacted households may lack running water, may perceive physical and mental health effects, and may experience financial or property impacts related to the drought. The reported drought impacts varied by county, possibly in part due to differences in geography, demographics, or the varied effects of work availability and property maintenance costs in Tulare and Mariposa Counties.

As a result of these CASPERs, Tulare County health officials conducted additional community outreach, particularly to address hygiene; explored the use of television as a messaging medium; and expanded behavioral health partnerships. Tulare County also used CASPER information to apply for grant funding to continue drought-related activities. The CASPER data prompted Mariposa County Public Health to recognize that some of their assumptions about residents' drought and climate perceptions were flawed (e.g., more residents would rely on government assistance in case of water shortages than originally believed). Health officials shared CASPER results with the Board of Supervisors, local Office of Emergency Services, and other local entities to drive forward actions and policies addressing the drought impacts. In the immediate term, health officials conducted outreach to promote graywater system usage (including incorporating a graywater question into a community event survey).

References

CDPH. (2015a). *Community Assessment for Public Health Emergency Response (CASPER) addressing the California drought – Mariposa County, California, November 2015.* Available from https://www.cdph.ca.gov/Programs/CCDPHP/DEODC/CDPH%20Document%20Library/Mariposa%202015%20CASPER%20report.pdf. CDPH-DEODC, 850 Marina Bay Parkway, P-3, Richmond, CA 94804.

CDPH. (2015b). *Community Assessment for Public Health Emergency Response (CASPER) addressing the California drought – Tulare County, California, October 2015.* Available from https://www.cdph.ca.gov/Programs/CCDPHP/DEODC/CDPH%20Document%20Library/Tulare%20CASPER%20report.pdf. CDPH-DEODC, 850 Marina Bay Parkway, P-3, Richmond, CA 94804.

Governor's Office of Emergency Services. (2015). *Drought Operation Report, September 9, 2015.*

Mariposa County Ordinance 809, (1991 September 3).

Office of the Governor. (2014). *Governor Brown declares drought State of Emergency.* Retrieved from https://www.gov.ca.gov/news.php?id=18368.

Office of the Governor. (2015). *Drought update, Wednesday, November 25, 2015.* Retrieved from http://drought.ca.gov/pdf/archive/WeeklyDroughtUpdate(11-25-15).pdf.

Stanke, C., Kerac, M., Prudhomme, C., Medlock, J., & Murray, V. (2013). Health effects of drought: Systematic review of the evidence. *PLoS Currents, 5.* http://dx.doi.org/10.1371/currents.dis.7a2cee9e980f91ad7697b570bcc4b004.

Tafoya, D. (2016). *Personal communication.*

CHAPTER

9

Methods: Questionnaire Development and Interviewing Techniques

Amy H. Schnall, Amy Wolkin, Nicole Nakata

Centers for Disease Control and Prevention, Atlanta, GA, United States

INTRODUCTION

Responding appropriately and effectively to the public health threats of disaster, whether natural or human-induced, requires timely and accurate information. Disaster epidemiologists use questionnaires to collect data and generate this information. The questionnaire and interview are essential components of data collection for epidemiologic studies. A well-structured and effective questionnaire is an objective means of collecting data from people affected by disaster about their health status, knowledge, beliefs and attitudes, and behaviors. Careful planning is crucial for systematic and thorough data collection. The researcher must know a priori the questions they want answered and must ensure the questionnaire measures what it is intended to measure. Developing a quality questionnaire is complex, and attention to length, format, and flow is essential. This chapter discusses several key steps and tips to developing an effective questionnaire and successfully completing an interview.

QUESTIONNAIRE DEVELOPMENT

Objectives

The first step to developing a questionnaire is to determine the objectives of the study or assessment. The objectives help determine the appropriate methodology (e.g., cluster sampling, stratified sampling, convenience sampling), mode (e.g., in-person, phone, web-based), and sampling frame (e.g., geographic location, specific population). The objectives will also help focus the extent of the questionnaire. Having concise objectives will guide

Disaster Epidemiology
http://dx.doi.org/10.1016/B978-0-12-809318-4.00013-7

question development and help keep the questionnaire short, without jeopardizing the reliability and essentiality of the data. Development of the objectives should include input from leaders and decision-makers. During a disaster response when the Incident Command System (ICS) or Incident Management System is activated, the discussion about objectives should include the incident commander along with the state epidemiologist and other key stakeholders. The incident commander will ensure that the objectives meet the needs of the response, fit within the larger response goals and objectives, and are not duplicative of other efforts. Strong objectives clearly describe the purpose of the project and guide the development of the questionnaire.

Data Collection Method

After the objectives are defined, the next step is to determine the appropriate data collection method and design. Several data collection methods can be used, each with their own advantages and disadvantages. Table 9.1 describes four common data collection methods and some of the advantages and disadvantages of each method. For example, in-person data collection provides an opportunity to hand out public health information, gives the community a chance to see that there are people who care about how the disaster has

TABLE 9.1 Advantages and Disadvantages of Different Data Collection Methods

	Advantages	Disadvantages
In-person	• Higher response rate • Relatively low cost • Ability to gain further detail/probe the respondent and visually aid with recall • Ability to hand out public health information regarding the disaster	• Small sample size compared with other methods • Small sampling frame • Potential interviewer bias
Mail	• Relatively low cost • Large sample size • Ability to have large sampling frame (even multi-state or jurisdiction) • No interviewer bias	• Potential of disrupted mail service during a disaster • Historically low response rates • Slow data collection: requires time to mail, respondent to complete, and mail back • Inability to gain further detail/probe the respondent • Inherent bias of respondent
Telephone	• Higher response rate • Ability to have a large sample size • Ability to have large sampling frame • Ability to gain further detail/probe the respondent • Fast data collection	• Potential of disrupted phone service during a disaster • Relatively high cost • Inherent distribution bias • Potential interviewer bias
Web	• Very low cost • Large sample size • No interviewer bias • Fast data collection	• Potential of disrupted Internet service during a disaster • Historically low response rates • May be difficult to advertise during a disaster • Inherent distribution bias

impacted them, and allows for the use of visual aids to facilitate recall; however, in-person data collection often requires more resources and time than other methods (Rothman et al., 2008). In contrast, web-based surveys are less resource intensive and can garner a large number of participants but are more likely to introduce selection bias (i.e., only those with internet access may participate) and can be difficult to administer during a disaster when there is loss of power or Internet access (Bowling, 2005; Schleyer & Forrest, 2000; Wyatt, 2000). The data collection method selected will have an impact on sample size and timing. For example, if the study design requires thousands of respondents, a web-based survey may be the most appropriate method.

Survey Methodology and Sampling Frame

In addition to the data collection method, the survey methodology will further define the questions. For example, if a Community Assessment for Public Health Emergency Response (CASPER) is the chosen survey methodology, all questions must be written to collect information at the household-level, as opposed to the individual, to be appropriate for a face-to-face interview (Bayleyegn et al., 2012). On the other hand, other survey methodologies may rely on mail and web-based surveys, which are self-administered. These questionnaires are not administered with an interviewer so should be written in plain English and easy for a respondent to understand and complete. Mail and web-based surveys allow for more response options to questions than interviews do, since the interview questionnaire is designed to sound natural when spoken out loud (Bowling, 2005; Schleyer & Forrest, 2000).

The sampling frame, which is the desired population from which the sample is drawn, will also have an impact on the questionnaire development. Not only will the sampling frame effect the decision of the type of data collection method due to the desired sample size and timing, the sampling frame may also impact the questions themselves. Questions should be targeted specifically toward the desired sampling frame. For example, if the population included in the sampling frame is only those persons or households affected by a "Do Not Use" water order, then questions should be specific to that injunction and not be so general as to apply to those unaffected.

Question Development

Once the objectives, data collection method, survey methodology, and sampling frame are defined, the questionnaire can be developed. Disasters are often chaotic and information about affected individuals, households, and communities is needed rapidly. The number of questions and types of responses (e.g., yes/no; multiple choice, open-ended) to each question can impact the length of the questionnaire and the amount of time until data are available to public health responders.

During a disaster response, organizations responding to the disaster have competing priorities and differing data needs. It is not possible to collect every piece of information to meet every need. Doing so would overwhelm the interviewer and interviewee. During a disaster, time is of the essence; a precise questionnaire will provide high-quality data in a short amount of time. A long questionnaire not only requires more time when administered but also subsequently requires more time for data entry, cleaning, and analysis. The goal is to have as few questions as possible to obtain the information required to meet the objectives. Again, concise objectives will help determine the critical information needed and therefore will guide development of questions to obtain that specific information. Peripheral questions and those that are "nice to know" should be avoided; questions should focus only on "need to know" information. Although it may be tempting to ask a lot of questions, the additional time required to develop and administer the questionnaire and to manage and analyze the data outweighs any potential benefit. Another tip for writing targeted questions is to think through the analysis. Drafting table shells for the results of the analysis helps ensure that each question is analyzable, interpretable, and provides the needed information.

It is helpful to use questions from existing data collection instruments or question banks when possible. This saves time and has the added benefits of assuring the reliability of the questions and the comparability with other data. Reliability is a measure of the extent to which a questionnaire produces consistent results (Shadish, Cook, & Campbell, 2002). There are several tests for reliability. Test–retest reliability measures repeatability, that is, the degree to which test results are consistent over time; and internal consistency reliability measures (e.g., Cronbach's alpha reliability coefficient, split-half analysis) determine how well items in the questionnaire measure the same construct or idea (Cronbach, 1951; Cozby, 2001). Unfortunately, during a disaster, time constraints do not allow for testing the reliability of new questions. Therefore, to increase the reliability of the questionnaire, include questions used previously in similar situations. The Centers for Disease Control and Prevention's (CDC) Disaster Epidemiology Community of Practice, houses disaster-related resources, including previously used questionnaires and question banks (CDC, n.d.). Similarly, the Council of State and Territorial Epidemiologists (CSTE) Disaster Epidemiology Subcommittee created a repository for disaster epidemiology tools and user guides available to the public (CSTE, n.d.). The National Institute of Environmental Health Sciences (NIEHS) Disaster Research Response Program offers disaster data collection tools and other resources to facilitate the organization and approval of disaster-related research studies (NIEHS, n.d.). Surveys that are not specifically designed for disaster studies can be modified for disasters or other public health emergencies, and include CDC's Behavioral Risk Factor Surveillance System, the Patient Health Questionnaire, and the Generalized Anxiety Disorder Screener (APA, n.d.; CDC, 2016; Spitzer, Kroenke, Williams, & Löwe, 2006).

Choosing open- versus closed-ended questions requires careful consideration. Open-ended questions do not have a list of responses to choose from; they allow respondents to give free-form answers (Bethlehem, 2009; Israel, Eng, Schulz, & Parker, 2005). Closed-ended questions limit responses, providing a list of choices from which they must choose to answer the question (Bethlehem, 2009; Israel et al., 2005). Although open-ended questions provide invaluable detailed information, they do not have a single, definitive answer. Response categories are unknown, require elaboration, and are difficult to standardize and analyze in a timely manner. Closed-ended questions, on the other hand, have a finite set of answers from which to choose and therefore produce results more timely. For this reason, closed-ended questions are used more often than open-ended to collect information during a disaster or emergency event because they are quicker to respond to and to analyze.

The wording of each question should also be considered for ease of administrating the questionnaire during an interview. The sample population, or audience, should always be taken into account. Tailor the questionnaire to the characteristics of the population, considering age, education level, familiarity with surveys, cultural bias, and language (Bethlehem, 2009; Rothman, Greenland, & Lash, 2008). Questions should be written using plain language techniques. They should be direct, use the simplest language possible, and avoid jargon and acronyms. It is beneficial to include any definitions necessary. For example, when asking questions about household preparedness, it is important to assess whether the household has an adequate supply of food and/or water. However, the meaning of "adequate" must be defined. Questions that present more than one concept are called "double-barreled" questions. These types of questions may confuse a respondent about how to answer. For example, asking a respondent "do you plan to move from the disaster area and buy a house within the next year" is a double-barreled question. This question presents two concepts, one about moving and one about purchasing a home and should be two separate questions. Vague questions should also be avoided. Be specific about the information needed. For example, when asking questions about drug use, instead of asking "do you use drugs?", ask "do you use illegal drugs?". This way the respondent knows if you are referring to prescription, over-the-counter, or illegal drugs. Finally, loaded or leading questions (i.e., those that suggest a particular answer) should be avoided as they can introduce bias into your data.

Pilot-test the questionnaire with other people who are similar to the sample population to ensure the question wording is appropriate and understandable by your audience. This will allow you to improve unclear questions or procedures and detect errors beforehand and, therefore, reduce systematic measurement error, which will improve the internal validity of the project (Bethlehem, 2009). It is ideal to conduct in-depth pilot field tests or focus groups; however, during a disaster there is rarely time for that level of detailed testing. At a minimum, pretesting the questionnaire with someone who was not involved in its development and, ideally, who is similar to the sample population, will demonstrate whether the questions are easily understood by the respondent, are answerable, and that the order of the questions and skip patterns are intuitive and easy to follow.

Questionnaire Design

When designing a questionnaire, there are many factors to consider. Questions should be arranged in a logical order, with like topics or response types (e.g., Likert scales, multiple choice) grouped together. Because the respondent may not complete the entire interview, it is important to have the most critical questions at the beginning of the questionnaire to ensure that information is captured. The questionnaire format should be easy to use and understand, for both the interviewer and the respondent. For example, font size should be legible, skip patterns, if any, should be clearly defined, and sections should be marked appropriately. Sensitive questions (e.g., income) are best placed toward the middle or end of the questionnaire, a point where the respondent has likely become more comfortable answering the questions and will not stop the interview because of the questions' sensitive nature.

The data collection mode, whether on paper or electronically, should be considered. Telephone and in-person interviews may use either paper or electronic questionnaires for data collection. The mode may impact the timing of the survey and the type of questions developed. There are limitations and benefits to each mode of data collection. Generally, while paper forms can be labor intensive in the data entry process (the "back end"), they require less time to develop (the "front end"), so field teams can begin data collection more quickly. However, paper forms may require a simpler questionnaire as complex skip patterns can be confusing to interviewers without the automation of an electronic form. In contrast, electronic forms require more work on the "front end" to develop and test the technology, and to train interviewers. The additional steps needed prior to implementing an electronic data collection form may delay data collection. Additionally, during a disaster, power outages and unreliable Internet connections may impact the ability for field teams to use electronic questionnaires.

Informed Consent

Prior to beginning an interview, the interviewer must obtain the informed consent of the respondent. Informed consent is the introduction to the questionnaire. A typical consent script will be brief, describe who the interviewer is (i.e., the agency they represent), why the project is being conducted, how long the interview will take, the anonymity or confidentiality of responses, the voluntary nature of participation, and an explicit request of consent (Bayleyegn et al., 2012). The consent should provide a phone number so that the respondent can contact the organization with any questions they may have at the time of the interview or in the future. Prior to implementing the questionnaire in the field, consult with the local regulations (e.g., area institutional review board, or IRB) to ensure that the consent script meets all requirements. The importance of the consent script, as the first introduction of the project to the potential respondent, cannot be overstated, as it may impact an individual's decision to participate. A well-written, concise consent script will likely lead to greater participation. Also, memorizing some of the consent script for a more natural delivery will greatly improve the rapport that the interviewer has with the potential respondent and can help encourage the respondent to participate in the interview. Consent may be either verbal or written, depending on your project and the local IRB requirements. Often, as with a CASPER, written consent is the only personally identifying information collected. In that case, verbal consent is allowable and preferred (Bayleyegn et al., 2012). If the interview is conducted in person, leave a copy of the consent script with the respondent for their future reference.

INTERVIEW TECHNIQUES

Creating a comfortable environment for the respondent is essential to having a successful interview. Prior to conducting the interview, every interviewer should practice and review the questionnaire to become comfortable with skip patterns, pronunciation of words, and meaning behind the questions. Familiarity with the consent and the questionnaire will increase the success of the interview. If the questionnaire is electronic, interviewers must be familiar with the technology and how to troubleshoot any issues that may arise. If possible, have a dedicated IT individual on call to help answer any technical problems.

During the interview, interviewers should be empathetic and respectful. In person, they should dress appropriately, maintain good eye contact, and not eat, drink, or chew gum during the interview. If working in pairs, one team member should give the interview while the other records the answers, which helps build rapport with the respondent. It may be helpful to determine which team member is the better interviewer, although this may change as the day progresses and the first interviewer tires. If a respondent is unsure about answering the more sensitive or emotionally challenging questions, remind the respondent that participation is voluntary and responses will be handled confidentially or anonymously, as applicable.

Standardization of Interviews

They key to successful interviews is standardization. Interview standardization involves asking the questions in the same order with the exact wording (Rothman et al., 2008). This increases reliability of the data and eliminates a source of bias and error. This is especially important when more than one interviewer is collecting data. The standardization process can be difficult because reading verbatim from the questionnaire may feel artificial or awkward for the interviewer. Although it may be tempting for interviewers to rephrase or paraphrase questions, doing so may change the question being asked. The questions must be read verbatim. When a respondent has difficulty with a question, the interviewer should pause and allow them time to answer. When a respondent needs clarification about a question, the interviewer should repeat the question first and then elaborate if needed using additional information provided in the script, to avoid leading the respondent to a specific answer.

In addition to standardizing the questions, interviewers must also standardize the answer choices. For close-ended questions, the interviewer will either read all the response options, or will let the respondent answer, and then select the appropriate response option(s) based on the answer. For open-ended questions, responses must be recorded in their entirety, precisely as the respondent gives them. Paraphrasing a respondent's answer may change the meaning. For example, if the respondent says that their household's greatest need is financial aid to help rebuild their damaged home, the interviewer should write that response down in full and not just write "money." If only "money" is recorded, it is unclear whether money is needed for repair, for clothes, in general, or for another reason. It is also essential that the interviewer does not "prefill" questions or finish the respondent's sentences, even if they believe they know the answer. If the interviewer thinks the respondent has answered a later question while answering the current question or in conversation, the interviewer must still ask the question and record the answer given at that time. This also applies to questions that may seem obvious. For example, if you are conducting an interview and the respondent answers the phone by saying their name or opens the door wearing a dress, even if the interviewer is confident the respondent is female, the interviewer still must ask if the respondent is male or female. A common question about household preparedness asked during CASPER interviews is whether or not the household has a pet. While some interviewers may feel awkward asking that question if the respondent is standing next to a dog, it is still essential to ask the question as the dog could be a neighbor's or a friend's pet that happens to be at the house at the time of the interview.

Completing the Interview

The questionnaire must be reviewed for completion prior to hanging up or leaving the household. If there are any incomplete answers or answers that need clarifying, ask the respondent immediately. Once the interviewer is confident that the questionnaire is complete, thank the respondent for their time and effort, let them know when results may be available and how they will receive them, and provide the respondent with relevant information to help them cope with the disaster (e.g., how to replace important documents that were lost/destroyed, where to find evacuation shelters).

CONCLUSION

The questionnaire and interview components may be the most important part of any study or assessment during a disaster. A well-written questionnaire can provide essential information for leaders and decision-makers to respond appropriately and effectively to the ensuing public health threats.

References

American Psychological Association (APA). (n.d.). Patient health questionnaire (PHQ-9 & PHQ-2). Retrieved from http://www.apa.org/pi/about/publications/caregivers/practice-settings/assessment/tools/patient-health.aspx.

Bayleyegn, T., Vagi, S., Schnall, A., Podgornik, M., Noe, R., & Wolkin, A. (2012). *Community Assessment for Public Health Emergency Response (CASPER) toolkit* (2nd ed).

Bethlehem, J. (2009). *Applied survey methods: A statistical perspective*. Hoboken (NJ): John Wiley and Sons.

Bowling, A. (2005). Mode of questionnaire administration can have serious side effects on data quality. *Journal of Public Health, 27*, 281–291.

Centers for Disease Control and Prevention (CDC). (2016). *Behavioral risk factor surveillance system*. Retrieved from http://www.cdc.gov/brfss/.

Centers for Disease Control and Prevention (CDC). (n.d.). Disaster epidemiology community of practice SharePoint site. Retrieved from https://partner.cdc.gov/DECoP.

Council of State and Territorial Epidemiologists (CSTE). (n.d.). Disaster epidemiology subcommittee: Tools repository. Retrieved from http://cste.site-ym.com/?DisasterEpiToolRep.

Cozby, P. C. (2001). *Measurement concepts. Methods in behavioral research* (7th ed). California: Mayfield Publishing Company.

Cronbach, L. (1951). Coefficient alpha and the internal structure of tests. *Psychomerika, 16*, 297–334.

Israel, B. A., Eng, E., Schulz, A. J., & Parker, E. A. (2005). *Methods in community-based participatory research for health*. San Francisco (CA): Jossey-Bass.

National Institute of Environmental Health and Sciences (NIEHS). (n.d.). NIH disaster research response (DR2): Data tools and resources. Retrieved from https://dr2.nlm.nih.gov/tools-resources.

Rothman, K. J., Greenland, S., & Lash, T. L. (2008). *Modern epidemiology* (3rd ed). Philadelphia (PA): Lippincott, Williams &Wilkins.

Schleyer, T. K. L., & Forrest, J. L. (2000). Methods for the design and administration of web-based surveys. *Journal of American Medical Informatics Association, 7*, 416–425.

Shadish, W. R., Cook, T. D., & Campbell, D. T. (2002). *Experimental and quasi-experimental designs for generalized causal inference*. Belmont (CA): Wadsworth, Cengage Learning.

Spitzer, R. L., Kroenke, K., Williams, J. B., & Löwe, B. (2006). A brief measure for assessing generalized anxiety disorder: The GAD-7. *Archives of Internal Medicine, 166*, 1092–1097.

Wyatt, J. C. (2000). When to use web-based surveys. *Journal of American Medical Informatics Association, 7*, 426–429.

Vignette: Investigating Foodborne Outbreaks

Marilyn Felkner, Venessa Cantu, Kevin McClaran

Texas Department of State Health Services, Austin, TX, United States

On November 21, 2014, the Texas Department of State Health Services (DSHS) received notification from the Centers for Disease Control and Prevention (CDC) that a case of listeriosis reported by Texas had been shown to be related based on pulsed field gel electrophoresis (PFGE, a type of DNA fingerprinting) to a nationwide cluster of listeriosis. Over the next 2 weeks, three additional Texas cases were identified as part of the cluster, as well as 28 other cases nationally.

Listeriosis cases are routinely interviewed using the CDC *Listeria* Initiative (LI) case form as soon as they are reported. The LI case form captures exposures to food items (or similar food items) that historically have been associated with listeriosis. Based on the number of cases exhibiting the cluster PFGE patterns and that no food item from the initial LI case form interviews was reported in high frequency, the CDC sent a hypothesis-generating questionnaire to states with cases to reinterview cases to expand on food items not captured on the LI form. The hypothesis-generating questionnaire includes hundreds of food items to identify any possible food that all or a majority of cases might have consumed. DSHS delegates interviews to the health department for the jurisdiction in which the case resides—a local or regional health department. The four Texas cases resided in two regions of the state. One region had three cases, one in a local health department's jurisdiction and two in the regional health department's jurisdiction. The interviews were conducted by the respective health departments.

The interview assigned to the local health department posed particular challenges. The case had died a few days prior to the interview necessitating interview of a surrogate. The still-grieving spouse could recall very little of the food history. The spouse did remember consumption of the caramel apples, and the epidemiologist conscientiously recorded this information even though it was not part of the case form. The regional health department epidemiologist aggregated and reviewed the three hypothesis-generating questionnaires prior to a CDC-requested joint CDC/DSHS interview of the cases. The regional epidemiologist noted that the hypothesis-generating questionnaire completed during interview by the local health department mentioned caramel apples and the national food retailer where the apples were

Disaster Epidemiology
http://dx.doi.org/10.1016/B978-0-12-809318-4.00014-9

purchased. During the first CDC/DSHS joint interview with another interviewee, the regional epidemiologist, ad hoc, asked about caramel apple consumption. The answer was affirmative.

The CDC notified other public health jurisdictions and other cases were queried about caramel apple consumption. CDC deployed a new supplemental questionnaire with additional questions focusing on caramel apple consumption. Within a few days, 15 of the 18 (83%) reinterviewed cases reported commercially produced, prepackaged caramel apple consumption. Additional questioning from the supplemental focused questionnaire revealed the source of the commercially produced, prepackaged caramel apples. Caramel apples were obtained and bacteria matching the clinical strains by both PFGE and whole genome sequencing (WGS) were identified from the apples. The apples were consequently withdrawn from the market.

At the conclusion of the investigation, 35 cases were reported by 12 states. Twenty-eight of the 31 (90%) cases interviewed reported commercially produced, prepackaged caramel apple consumption, a novel food vehicle for listeriosis.

This successful investigation illustrates a four important components of surveillance—the molecular laboratory, the use of hypothesis-generating questionnaires, the astute epidemiologist, and collaboration between public health entities when conducting disease outbreak investigation.

While classic point source foodborne illness outbreaks—the fund-raiser barbecue, the catered wedding reception—still occur, the current global food distribution network calls for more sophisticated tools. In the past two decades, the molecular laboratory has evolved to fill this need. Being able to identify specific strains within hundreds of serotypes has led to the descriptive analogy "molecular fingerprinting." While characteristics of the bacteria such as mutability and clonality can limit the utility of PFGE, WSG holds promise of even more definitive identification of related organisms.

To maximize the use of this tool, the CDC operates PulseNet, a database of PFGE and, now, WGS patterns from across the nation. Clinical isolates from ill individuals diagnosed with *Salmonella, Listeria,* and Shiga toxin–producing *Escherichia coli,* undergo molecular testing in public health laboratories throughout the United States, and the data are consolidated at the CDC. Additionally, state and federal regulatory agencies such as the Food and Drug Administration (FDA) and United States Department of Agriculture (USDA) submit specimens obtained from foods or food production facilities. Using sophisticated algorithms that compare the number of submissions to seasonally adjusted baseline data, the CDC identifies significant increases in the number of submissions of particular strains. When a threshold is surpassed, the CDC alerts affected states and requests investigations using the hypothesis-generating questionnaire. In the example of the caramel apples, what once would have been seen as one or two isolated cases of *Salmonella* in communities across the country with no chance of reaching a significant odds ratio can be viewed as links in a large network. Finding an identical bacterial strain in the suspected food or other vehicle provides near certainty in the cause of the outbreak and makes removing the contaminated product from the market possible, protecting both consumers and industry.

The hypothesis-generating questionnaire is also an important adaptation necessitated by global food distribution. While in point source foodborne illness outbreaks, a finite menu is usually available, this is not the case in outbreaks stemming from the global food market. The hypothesis-generating questionnaire, also known as the shotgun questionnaire, must try to capture any food that the person might have eaten during the pathogen's incubation period. Giving a selection of foods to choose from is more likely to elicit a variety of responses than asking people to list foods they have eaten, although allowing "free text" responses is also important as the caramel apple example illustrates. Once a general category or type of food has been identified, additional focused questionnaires can be developed and administered, which include more nuanced questions about the product such as preparation method, brand, flavor, or store/restaurant of origin.

Ultimately, molecular tools and questionnaires are only as useful as the professionals who use them. Administering questionnaires requires numerous abilities including perseverance, persuasiveness, courtesy, sensitivity, logic, attention to detail, and multitasking. The epidemiologist must persevere in reaching the case, often during nonbusiness hours, and then be sufficiently persuasive to convince the case, who is usually feeling well by this time, to remember details about an unpleasant experience. Furthermore, the epidemiologist must remain courteous—even if the case is apathetic about the need for such questions or confrontational about government intrusion—and keep the case engaged for 45 min or more to complete the interview. In other situations such as interviewing a person who is gravely ill or a surrogate who has experienced the loss of a loved one, an interviewer must exercise sensitivity to the situation. While asking the questions, the epidemiologist must acknowledge and document information volunteered by the interviewee even when not specifically requested as did the interviewer in the caramel apple outbreak. The interviewer must also be alert to inconsistencies in the case's response and clarify as these occur (e.g., I do not eat peppers. I do eat jalapenos.). Furthermore, the epidemiologist must accurately record the responses in a way that will be legible and understandable for data entry and future analyses.

Solving cross-jurisdictional outbreaks like the one described here requires the collective effort of all jurisdictions involved. Without communication and collaboration by public health at the local, regional, state, and federal levels, the source of this outbreak may never have been identified. While epidemiologic investigations have benefitted dramatically from improved technology, effective investigations continue to rely on data collected by knowledgeable, collaborative epidemiologists.

CHAPTER

10

Applications: Social Vulnerability to Disaster (Hampton and Hertford Counties—Isabel)

Danielle Spurlock
University of North Carolina, Chapel Hill, NC, United States

INTRODUCTION

Every community faces risks posed by natural hazards (e.g., floods, hurricanes, ice storms, wildfires, and earthquakes) and technological hazards (e.g., chemical spills or explosions). However, the likelihood a hazard becomes a disaster is affected by the spatial and social structure of a community. While physical vulnerability to a hazard can result from sociopolitical decisions that place residential and commercial structures in hazard-prone areas, social vulnerability identifies characteristics such as income, age, disability status, availability of transportation, and race/ethnicity as factors contributing to a population's vulnerability to negative outcomes before, during, and after a hazard event (Cutter, Boruff, & Shirley, 2003; Morrow, 1999; Van Zandt et al., 2012). For example, Hurricanes Katrina and Rita highlighted the challenges that individuals and households faced in evacuation, sheltering, and prolonged recovery due to extended periods of unemployment, a complicated federal aid process, the dissolution of social networks, and inadequate resources to repair or replace damaged homes and cars (Laska & Morrow, 2006).

Initiated 3 months prior to Hurricane Katrina's landfall, a three year Federal Emergency Management Agency (FEMA)-funded Emergency Preparedness Demonstration Project (EPDP) focused on the six states (Delaware, Maryland, North Carolina, Pennsylvania, Virginia, and West Virginia) affected by Hurricane Isabel, as well as the District of Columbia. The project's ultimate goal was to develop a model program to improve awareness and preparedness in socially vulnerable communities. This chapter examines how two communities (Hertford County, North Carolina and Hampton, Virginia) assessed their social vulnerability. It begins with profiles of the communities' physical and social vulnerability and recounts how the planning process engaged and empowered local residents. Next, it describes the

Disaster Epidemiology
http://dx.doi.org/10.1016/B978-0-12-809318-4.00015-0

community-driven creation, testing, and administration of survey and interview instruments, as well as results from the communities' data collection efforts.

PHYSICAL AND SOCIAL VULNERABILITY IN HERTFORD COUNTY

Located in the northeastern region of North Carolina, Hertford County is a rural, inland county subject to riverine flooding, storm surge, and wind-induced power outages. Between 1995 and 2003, Hertford experienced seven flooding events, with the last three events (Hurricane Floyd, September 1999; Tropical Storm Allison, June 2001; and Hurricane Isabel, September 2003) receiving federal disaster declarations. Although the Hertford County Emergency Management Office (EMO) oversaw the purchase and demolition of 101 flood-prone residential structures through the Hazard Mitigation Grant Program in the aftermath of Tropical Storm Allison, a number of residential and commercial properties remained at risk. Limited data on hazard-prone areas also complicated the emergency planning process; when Hertford created their 2004 Hazard Mitigation Plan, only 1978 floodplain maps were available and the county only had digital parcel, zoning, and tax data for a few towns.

At the time of the project, Hertford exceeded state and national percentages on a number of factors associated with social vulnerability. The Department of Commerce ranked Hertford among the most economically distressed counties in North Carolina, with a median household income of $26,422, well below the US ($48,451) and North Carolina ($42,625) figures. Poverty rates in Hertford (18.3%) exceeded the percentage observed in North Carolina (12.3%) and the United States (12.4%) with more children and elderly living below the poverty line (21% for both groups) compared with the rest of North Carolina's children (15.7%) and elderly (13.2%). The Hertford population aged five and over with a reported disability (31.9%) exceeded state (21.1%) and national (19.3%) percentages. The percentage of Hertford households without access to vehicles (13.1%) also exceeded the percentages of North Carolinian households (7.5%) and US households (10.3%). Finally, approximately 60% of the county's population was African American—one of the highest percentages in the state, which had implications for household access to insurance (Peacock & Girard, 2000; Tierney, 2006).

Hertford County presented a rural example of social vulnerability coupled with staff and resource limitations that hindered the local EMO's ability to engage socially vulnerable populations in emergency preparedness planning. Hertford County's EMO was largely the responsibility of a single individual who also oversaw Emergency Medical Services (EMS). Like many low-wealth, rural counties, Hertford County hired an outside consulting firm to prepare its hazard mitigation plan. While this option helped the county comply with the Robert T. Stafford Disaster Relief and Emergency Assistance Act, the resulting plan was not tailored to Hertford County. Additionally, although County residents had an opportunity to voice their concerns and comment on the plan in two community meetings, the main advertisement for these meetings was a 1/16th page ad in the classified section of the Roanoke-Chowan News-Herald, a local newspaper with a circulation of 7800 in a four-county area. This type of advertisement satisfies the stipulations of North Carolina state law about public participation, but the mode and circulation limits the effectiveness of this communication mode.

PHYSICAL AND SOCIAL VULNERABILITY IN THE CITY OF HAMPTON

The City of Hampton is the oldest continuously settled, English-speaking community in the United States, and was the site for the first free public schools in the United States. Bordered by the Chesapeake Bay and Hampton Roads, Hampton's location on the Lower Peninsula of Virginia increases its susceptibility to a number of natural hazards including hurricanes, coastal flooding, storm surge, flash flooding, and tornados. At the time of the project, Hampton had suffered from eight significant flood events since 1999, including Hurricane Floyd, Hurricane Irene, Hurricane Isabel, and Tropical Storm Gaston. Hurricane Isabel, the impetus for the EPDP, was one of most significant events to hit Virginia since Hurricane Hazel in 1954. The 2006 Peninsula Multi-Jurisdictional Natural Hazards Mitigation Plan estimated damages from Hurricane Isabel at $925 million.

According to the 2005–07 American Community Survey, poverty rates and median household income in Hampton suggest at least 13% of households would face obstacles with the financial aspects of emergency preparedness such as purchasing supplies. The median household income in the City of Hampton in 2006 was $47,408, about $2600 lower than the US average ($50,007), and approximately $11,000 less than the Virginia median household income. The poverty rate of 13.2% in Hampton exceeded the rate observed in Virginia (9.9%) and was comparable to the United States as a whole (12.4%). Approximately 20% of children and 9% of elderly living in Hampton had an income below the poverty line in the prior 12 months. Further, 14.6% of the population aged 5 and older lived with a reported disability while 40.6% of the population aged 65 and older lived with a disability. Depending on an individual's disability status, they may have difficulty sheltering-in-place, local public shelters may be unable to accommodate their medical equipment or care needs, or they may need a more complicated evacuation plan (i.e., organizing transportation with equipment and care in a geographically distant jurisdiction).

Prior to the initiation of the EPDP, Hampton residents identified the proximity of low-income populations to storm surge and the necessity for evacuation for even a Category One hurricane as an emergency planning concern. Additionally, negative consequences such as disruption of electricity, limited transportation options to shelters and supply distribution sites, and ineffective communication via traditional media such as newspapers and radio reports before and after an event were felt more acutely by low income residents. The City of Hampton presented an urban example of social vulnerability within a community with a reputation for well-integrated local and regional organizations and long-standing efforts to engage residents in civic life.

EMERGENCY PREPAREDNESS DEMONSTRATION PROJECT AND THE DATA COLLECTION PROCESS

During an 18-month period between 2005 and 2007, the City of Hampton, Virginia, and Hertford County, North Carolina, collaborated with MDC, Inc., a private nonprofit, and the Center for Urban and Regional Studies (CURS) at the University of North Carolina at Chapel Hill to engage in a four-phased process aimed at engaging and empowering local residents: Community Entrée and Building Connections; Constructing the Current Reality; Data to Action; and Implementation. The data collection processes described below reflect efforts completed during on the Community Entrée and Constructing the Current Reality phases and detail the creation of survey and interview instruments that reflected the data needs and collection methods designed by rural and urban Emergency Preparedness Demonstration Team (EPDT) members.

The first critical step was to engage emergency preparedness stakeholders. In Hertford, initial contact was made with county government leaders and emergency management officials to introduce the project and gain their approval. As important gatekeepers to the local governance structure, these agencies were knowledgeable of existing vulnerabilities, plans, and the constraints on governmental emergency management. These partners also provided an important link between the project and the community organizations directly and indirectly involved in emergency preparedness and were necessary participants to ensure project findings were incorporated into the official emergency mitigation and response plans.

Roanoke Economic Development Inc. (REDI), a recognized and respected community-based organization, acted as the local host organization. REDI considered the project's goal of addressing barriers to emergency preparedness and awareness in socially vulnerable communities as complimentary to their goals of community and economic development. Staff from REDI and the Hertford County Cooperative Extension helped create a list of influential community leaders and organizations, and staff members with long-time ties in the community made initial contact. Finally, the project employed a "community coach" to act as a facilitator at all of the community meetings. The community coach was skilled in engagement and capacity building and had specific experience working with low-wealth communities. Her role was to build cohesion among team members and to guide them through the planning process.

In 1994, the Hampton City Council created the Neighborhood Office to lead and staff the Neighborhood Initiative, which consisted of three programs: (1) Neighborhood College, (2) Leadership Institute, and (3) Seminar Series. The Neighborhood College aimed to improve civic awareness by providing courses on how government operates, homeownership, and neighborhood organizing. The Leadership Institute focused on building skills around volunteer recruitment and retention, and effective meeting facilitation. The Seminar Series covered topics such as dispute resolution and diversity training. The presence of the Neighborhood Initiative contributed to a population of residents with substantial capacity in partnership creation, neighborhood organizing, and planning processes. While the Neighborhood Office was not well versed in emergency preparedness, they occupied an important role as liaison between local government, regional agencies, neighborhood organizations, and the residents and were assigned the role as local host organization.

The Constructing the Current Reality phase consisted of a self-assessment process where the EPDT gathered information and constructed a shared understanding of physical and social vulnerabilities. Participants shared personal disaster experiences, including accounts of Hurricane Isabel, heard presentations about emergency management, and joined in facilitated discussions about the challenges facing socially vulnerable populations. Service providers in both communities lamented the lack the staff time or resources to incorporate effective communication strategies.

> Even though we are putting information on the website and in the newspaper, a lot of individuals do not see it. And, for example, the elderly population was not brought up with today's technology so they do not go to the website. And everyone is becoming technologically more conscious thinking they can reach the vast majority of people, but you are missing the population that needs your help the most.

In turn, residents cited the inflexibility of the current system of preparedness, response, and recovery as a concern, as well as the difficulty of understanding vague messages that were not provided in a timely fashion. In both communities, a community survey was identified as a method of gathering information from a broad group of community residents.

DATA COLLECTION, ANALYSIS, AND FINDINGS FOR HERTFORD COUNTY

As part of the original project design, researchers at CURS proposed a randomized, door-to-door survey of Hertford residents to gather information about preparedness. However, EPDT members altered this approach. Team members questioned whether the recommended methodology (letter, call, door-to-door) would result in good participation because of limited literacy. Furthermore, many individuals in the community were distrustful of the research process after a history of extractive research where research projects were conducted, but research findings were never shared with the community. Instead, EPDT team members suggested (1) pairing a community member with CURS staff for interviews to gain entrée into local households, (2) collaborating with community stakeholders to generate an initial list and snowball selection of participants for subsequent interviews, and (3) intercepting residents in settings such as the county's social services office. The interview guide covered topics such as household preparedness for emergencies, the receipt of information such as preparedness tips and evacuation warnings, and barriers to better emergency preparedness. The reconstructed process resulted in 39 semistructured interviews completed over a week's time in Hertford County by three people from CURS who were each paired with a Hertford County resident. The interviews resulted in four key findings.

1. Interviewees indicated that many socially vulnerable households in Hertford County already take preparedness actions for themselves (i.e., gathering supplies, removing yard items that could become flying debris, and filling up their cars with gasoline) and their neighborhoods (i.e., communicating with family and friends about the possible events and purchasing supplies for elderly neighbors) although these activities were not always readily identified as "emergency preparedness."
2. Interviewees raised concerns about financial limitations on the preparedness actions they could take. For example, a household might take preparedness actions such as stockpiling food and water, but is unable to purchase enough. "No matter what you do, though, these actions are only effective for one or two days. If the storm comes, nine times out of ten, you don't have enough money to get all the things you need."
3. Interviewees stated that some households were unaware of impending storms and, thus, did not take preparedness actions. "A lot of people don't, in this area, do not look at or read newspapers. A lot of people don't look at TV, you know, not at the news and weather channels."

4. Interviewees repeatedly mentioned that if any individual lacked experience with emergencies or believed that the risk of negative outcomes was low, it diminished the urgency to take preparedness actions. "Isabel, even though they told us that it was coming, I don't think the people or the public really thought that it was going to be as serious as it was because they hadn't had any serious ones before." Other interviewees felt that even with prior experience, residents would not make preparations because "once the storm passed, people have the tendency to say, 'We made it through, and we did okay.'"

These findings were presented to the EPDT, individuals who have been interviewed, and other invited guests including religious and community leaders who were not yet involved in the project. In addition to the findings from the interviews, the following emergency preparedness challenges were articulated by EPDT members during their planning meetings.

1. *Communication* with the non–English-speaking population of migrant workers and residents was difficult as they did not receive print, radio, or televised preparedness messages.
2. *Transportation to designated shelters* for people who do not have cars or cannot drive because of a disability or other limitation was unavailable.
3. *Transportation after an event* when roads may be impassable because of debris from a storm, such as in Isabel, making it impossible to get from point A to point B.
4. *Isolation* can be temporary or longer depending on the circumstances such as language barriers, location within a rural area, and/or elderly or individuals with a disability being homebound.
5. *The sheltering and caring for special needs populations* who may need medical care or machines that require electricity to operate and sustain well-being.

Based on the challenges described by the household interviews and the issues elucidated during facilitated discussions, the Hertford County EPDT developed the following vision statement for their work in July and August of 2006.

Hertford County EPDP Vision Statement

All citizens of Hertford County, including those who are socially vulnerable (i.e., disabled and/or economically or socially disadvantaged), will have the information, supplies, and/or transportation they need to survive a disaster without suffering physical harm or deprivation of basic needs: water, shelter, and medical supplies. Organizations of all kinds will work together to ensure that all citizen are adequately prepared and will lead a coordinated response and recovery effort in the wake of a disaster that will be inclusive of all citizens.

This vision articulated the overarching tasks of the EPDT and offered a benchmark to measure whether subsequent goals and strategies were in line with the hopes and aspirations of the team. When reflecting on the planning process, Hertford EPDT members referred to the group's common vision for improving Hertford County as an essential condition for service providers, government officials, and residents to work together. As part of their public awareness campaign, the team designed and distributed 10,000 magnets with checklists delineating what households should have on hand to prepare for an emergency. The checklist was broken down into weekly lists of items that could be purchased on a limited budget.

DATA COLLECTION, ANALYSIS, AND FINDINGS FOR CITY OF HAMPTON

The key step in mobilizing the Hampton EPDT was the provision of an opportunity for service providers and residents to discuss previous emergency experiences separately before coming together to provide constructive criticisms about past responses. The services providers and community members undertook separate but parallel information gathering processes to meet the unique needs of both groups. Service providers met during business hours and discussed challenges specific to their agencies and organizations. Community members held evening meetings with presentations and discussions aimed at increasing awareness local emergency preparedness procedures and to share their experiences with local service providers prior to joint meetings. The presence of the skilled community coach allowed all participants to contribute to meetings where successes, criticisms, and challenges could be provided in a constructive environment. With the existing community engagement experience and planning capacity among key stakeholders in the EPDT, the community coach was able to focus on balancing participation from all parties at meetings and to reflect the key decisions made throughout the process back to the planning group to help build consensus. One participant stated that the coach was able to pose the questions that would be difficult to ask diplomatically without risking damage to well-established relationships.

To provide the Hampton EPDT with more information about residents' emergency awareness and preparedness activities, the core resident planning team constructed a survey instrument that covered knowledge and interaction with emergency management service providers, the receipt and usefulness of emergency preparedness materials, and household preparedness activities including evacuation and sheltering plans. The team then coordinated with 45 students from Hampton University to conduct a door-to-door survey in three Hampton neighborhoods: Old North Hampton, Tyler-Seldendale, and Wythe-Phenix. Residents of a fourth neighborhood, Lincoln Park, conducted a door-to-door survey of their community in early April 2007. In total, 189 interviews were completed and analyzed in the three neighborhoods. On hearing about the survey, Lincoln Park community residents self-organized and conducted 130 door-to-door interviews. Key findings included the following:

- Over 48% of households in each neighborhood surveyed did not have an emergency disaster plan.
- At least 37% of households in each neighborhood surveyed did not have an evacuation plan.
- More than half of surveyed households in each neighborhood did not have a disaster supply kit.
- One-fifth of surveyed households in each neighborhood had at least one member living with a disability.
- In Lincoln Park, only 25% of the surveyed households have access to a personal vehicle.
- Nearly half of surveyed households in each neighborhood did not know where to go if they needed to use a public shelter.
- Nearly three-quarters of households in each neighborhood surveyed were not aware of any specific plans for evacuating and/or sheltering the sick, elderly, children, or others who might have difficulty taking care of themselves.

The EPDP team used these data and their facilitated discussions to identify four key focus areas: (1) community awareness of household preparedness including evacuation and sheltering plans, (2) methods to combat household preparedness apathy and denial, (3) lack of connection between of the city's Emergency Management Bureau and residents, and (4) dissemination of emergency preparedness, sheltering, and evacuation information. Based on survey and prioritization criteria (i.e., *effectiveness*, *applicable* to all four communities, and *feasible* given the short timeline of the project), the Hampton EPDT selected three strategies:

1. Use preexisting materials as a guide in creating Hampton-specific materials about preparedness, sheltering, evacuation, and long-term recovery.
2. Create new communication pathways to stimulate individuals not engaged by current campaign and products.
3. Distribute materials using methods determined to be most appropriate, credible, and instructive for each EPDP-targeted neighborhood area.

CONCLUSION

The main limitation of the data collection methods outlined here is generalizability. With any survey or interview, those who choose not to participate represent important but unknown data because their reasons for declining could have some bearing on the data collected. For example, if lack of trust in an outside organization leads a person to decline to take part in an interview, this decision could also affect how they approach emergency preparedness messages or other programming. Despite these shortcomings, both of these examples illustrate how a data collection process can incorporate local knowledge, particularly that of socially vulnerable populations, in a meaningful way. The process provided community-specific data from residents, government officials, and service providers, which enabled them to be integral members in the explanation of preparedness behavior in their community. Most importantly, representatives of socially vulnerable populations were not just the subject of the project but engaged participants with the ability to influence the structure and implementation of the data collection and final projects focused on their communities.

References

Cutter, S. L., Boruff, B. J., & Shirley, W. L. (2003). Social vulnerability to environmental hazards. *Social Science Quarterly, 84*(2), 242–261.

Laska, S., & Morrow, B. H. (2006). Social vulnerabilities and Hurricane Katrina: An unnatural disaster in new orleans. *Marine Technology Society Journal, 40*(4).

Morrow, B. H. (1999). Identifying and mapping community vulnerability. *Disasters, 23*(1), 1–18.

Peacock, W. G., & Girard, C. (2000). Ethnic and racial inequalities in hurricane damage and insurance settlements. In W. G. Peacock, B. H. Morrow, & H. Gladwin (Eds.), *Hurricane andrew: Ethnicity, gender, and the sociology of disasters* (pp. 171–190). Miami, FL: Laboratory for Social and Behavioral Research, Florida International University.

Tierney, K. (2006). Social inequality, hazards, and disasters. In R. Daniels, D. Kettl, & H. Kunreuther (Eds.), *On risk and disaster: Lessons from Hurricane Katrina* (pp. 109–128). University of Pennsylvania Press.

Van Zandt, S., Peacock, W. G., Henry, D. W., Grover, H., Highfield, W. E., & Brody, S. D. (2012). Mapping social vulnerability to enhance housing and neighborhood resilience. *Housing Policy Debate, 22*(1), 29–55.

CHAPTER

11

Applications: Emergency Responder Health Monitoring and Surveillance

Successful Application

Renée H. Funk

Centers for Disease Control and Prevention, Atlanta, GA, United States

BACKGROUND

Emergency responders are valuable to our society and need to be protected so that they can go home safely each night and are able to respond again tomorrow. The response to the World Trade Center disaster demonstrated that there were significant gaps and deficiencies in health monitoring and surveillance of emergency response workers (including police, fire, and emergency medical personnel, as well as public health personnel and cleanup/repair/restoration/recovery workers). These gaps and deficiencies were documented in the RAND reports (NIOSH, 2004) prepared following the World Trade Center response and by others (Herbert et al., 2006; Newman, 2014; Reissman et al., 2012), but they persisted and were observed again in the response to and recovery from Hurricanes Katrina and Rita (Bergan et al., 2015; CDC, 2006; Rusiecki et al., 2014) and the Deepwater Horizon oil spill (Kitt et al., 2011; Lowe et al., 2015).

The persistence of these gaps and deficiencies in emergency responder health monitoring and surveillance (ERHMS), despite considerable attempts to anticipate and correct them, emphasized the need for a coherent, comprehensive approach to protecting these groups of workers and for detailed, practical guidance on implementing such an approach. Any effort to meet this need had to incorporate a variety of measures, including the following: (1) medical screening that focuses on assessment of fitness and ability to safely and effectively deploy on a response; (2) training regarding hazards to be anticipated and protective measures to mitigate them; (3) approaches to centralized tracking or rostering of responders; (4) surveillance and monitoring for exposures and adverse health effects, including supporting efforts in environmental monitoring and assessment; (5) out-processing assessments on completion of response duties and deployments; and (6) follow-up or long-term surveillance

Disaster Epidemiology
http://dx.doi.org/10.1016/B978-0-12-809318-4.00016-2

or monitoring for potential delayed or long-term adverse effects of the deployment experience. Similarly, such a system needed to include activities to be performed at all stages in the response spectrum—prior to, during, and following deployment. Any guidelines or recommendations for procedures to implement these protections had to be fully compatible with and function within the National Incident Management System structures, which have been adopted as the accepted standard organizational focus for emergency response at all levels (local, state, and federal) and for all incident sizes and types. Furthermore, the procedures needed to be understandable and usable by Incident Command System leadership and health, safety, and medical personnel.

In response to this continuing need, a consortium of federal agencies, state health departments, and volunteer responder groups was convened by the National Institute for Occupational Safety and Health (NIOSH). The product of those deliberations became a set of guidelines and recommendations called "Emergency Responder Health Monitoring and Surveillance (ERHMS): Technical Assistance Document (TAD)" and "Emergency Responder Health Monitoring and Surveillance (ERHMS): A Guide for Key Decision Makers." These guidelines were endorsed by and ultimately published by the National Response Team, Worker Safety and Health Subcommittee (National Response Team, 2012a, 2012b).

EMERGENCY RESPONDER HEALTH MONITORING AND SURVEILLANCE

The ERHMS system is a framework that includes recommendations and tools specific to the protection of emergency responders during all phases of a response, including predeployment, deployment, and postdeployment phases (Fig. 11.1). ERHMS addresses all aspects of protecting emergency responders and is applicable over the full range of emergency types and settings. It is intended to be of use to all those involved in the deployment and protection of emergency responders, including incident management leadership; leadership of response organizations; health, safety, and medical personnel; and the workers themselves.

The intent of ERHMS is to:

- identify exposures and/or signs and symptoms early in the course of an emergency response
- prevent or mitigate adverse physical and psychological outcomes
- ensure workers maintain their ability to respond effectively and are not harmed during response work
- evaluate protective measures
- identify responders for medical referral and possible enrollment in a long-term health surveillance program

The topics covered in each of the phases are:

Predeployment phase:

- Rostering and credentialing of emergency response and recovery workers
- Health screening for emergency responders

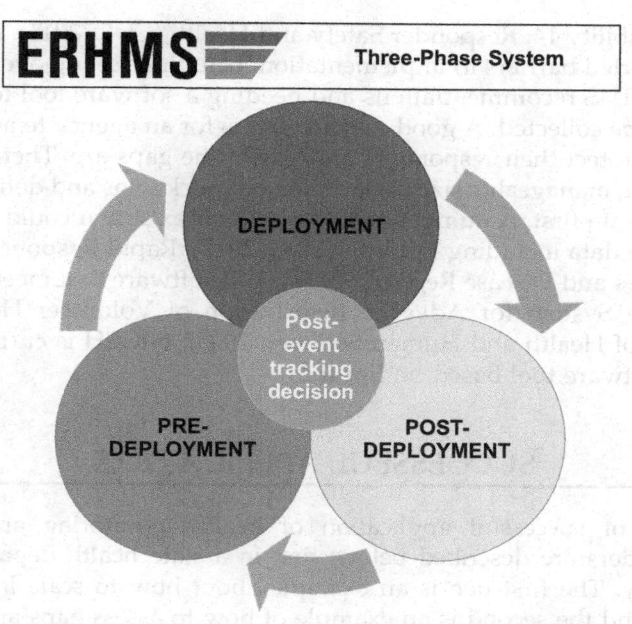

FIGURE 11.1 The three phases of the Emergency Responder Health Monitoring and Surveillance system.

- Health and safety training
- Data management and information security

Deployment phase:

- On-site responder in-processing
- Health monitoring and surveillance during response operations
- Integration of exposure assessment, responder activity documentation, and control
- Communications of exposure and health monitoring and surveillance data during an emergency response

Postdeployment phase:

- Responders out-processing assessment
- Postevent tracking of emergency responder health and function
- Lessons-learned and after-action assessments

Online training and additional ERHMS resources are available on the NIOSH website at: https://www.cdc.gov/niosh/topics/erhms/.

GETTING STARTED

A number of emergency response agencies have begun to use and implement the ERHMS guidelines to protect their responders. For example, it has been useful to state and local public health agencies in meeting their requirements for Public Health

Preparedness Capability 14: Responder Safety and Health (CDC, 2011). However, response agencies have reported barriers to implementation including feeling overwhelmed with the breadth of the ERHMS recommendations and needing a software tool to manage the large amount of data to be collected. A good way to start is for an agency to assess what they are already doing to protect their responders and where the gaps are. There are ways to scale implementation to a manageable size by focusing on quick wins and defining a subgroup of responders to focus on first. A number of software tools exist that could be used to manage all or parts of these data including Epi Info (CDC, 2017), Rapid Response Registry (Agency for Toxic Substances and Disease Registry, 2006), and software that meets the requirements for the Emergency System for Advance Registration of Volunteer Health Professionals (U.S. Department of Health and Human Services, 2005). NIOSH is currently developing a comprehensive software tool based on Epi Info.

SUCCESSFUL APPLICATION

Two examples of successful application of health monitoring and surveillance for emergency responders are described below; first in a state health department and second in a federal agency. The first one is an example about how to scale implementation to a manageable size, and the second is an example of how to assess gaps and fill them.

Georgia Department of Public Health

The Georgia Department of Public Health (GA DPH) quickly and successfully developed an online system to conduct active monitoring of all travelers returning from the three affected West African countries during the 2014 Ebola response. Their existing State Electronic Notifiable Disease Surveillance System was the platform (GA DPH, 2017). While the Ebola monitoring system was focused on the general public, they wanted to adapt the system to more fully address the ERHMS framework for their own workforce and named this new system, Responder Safety, Tracking, and Resilience (R-STaR). It would also help them meet Public Health Preparedness Capability 14, which focuses on protecting public health staff responding to an incident.

During October 2016, Hurricane Matthew was approaching, and GA DPH leadership knew that staff would be spread over Eastern Georgia. Leadership wanted to be able to monitor staff during the hurricane response. They adapted R-STaR within 24 h, and the GA DPH Emergency Operations Center (EOC) section leads sent out surveys to their responders to self-register in the system (Fig. 11.2). Responders reported that R-STaR was easy to use despite not having any training on it. After their profile was created through self-registration, staff received an email each day with a link to a survey for them to document their deployment activities for that day as well as complete a health and safety check (Fig. 11.3). This survey took approximately 1 min each day. Once the responders completed the surveys, the staff in the EOC were able to query R-STaR about an individual responder or run a report about all or a subset of responders.

Responder Registration Email

Hundreds of responders will go into the field in response to Hurricane Matthew. In order to ensure your health and safety, it is critical that we track your deployment activities to identify injury or illness during or after your deployment. Follow the below instructions to track your deployment activities:

Please complete the survey at this link to create a responder profile:
https://sendss.state.ga.us/sendss/!dynamicsurvey.surveypublicprompt?pQATemplateId=4318

Once this profile is created you will receive an email each day that will allow you to document your deployment activities, this will take less than a minute to complete each day.

If you have already been deployed, you can enter deployment activities that have already occurred when you create your responder profile.

We Protect Lives.

FIGURE 11.2 Responder registration message distributed by the Georgia Department of Public Health Emergency Operations Center in advance of Hurricane Matthew.

Daily Deployment Health and Safety Check

Health and Safety Check			
Please select your name from the list, then provide your current status and any updates on your location and health status			
1. Please select your name:	Doe, Jane ▼		
2. Current Status:	No Longer Deployed ▼		
3. Date of Safety Check	11 11 2016		
4. Deployment Roles and Duties	Epi, EOC		
5. Deployment County (if more than one, enter primary here and indicate others below)	Fulton ▼		
6. Length of your deployment in hours:	8		
7. Deployment Location (City, Facility, etc.)	2 PT		
Exposure / Injury Information			
8. Potential Exposures	Yes	No	Unknown
9. Injuries Sustained	Yes	No	Unknown
10. Notes (please record any exposures, injuries, or other items of note)	Epi, EOC		
SAVE			

FIGURE 11.3 Daily deployment health and safety check survey distributed to Georgia Department of Public Health staff.

Ultimately, 128 responders registered and reported from 11 health districts and more than 30 counties. This included staff from emergency medical services; environmental health; epidemiology; immunizations; nursing; women, infants, and children; and others sections. Six responders reported potential exposures to blood and body fluids, and each one was contacted to determine their health status and if additional medical follow-up was needed.

GA DPH evaluated this trial run of R-STaR and found that it was well received. Responders found it easy to use despite the last minute roll-out and lack of training. Responders liked knowing that someone was looking out for them, and supervisors liked knowing where their responders were, what they were doing, and that they were safe. In the past, GA DPH EOC has struggled to track the resources being used during a response and saw that R-STaR enabled them to accurately track human resources for the first time. Overall, GA DPH staff learned a lot from this trial run, and the lessons learned will be incorporated into the next version of R-STaR.

This serves as an excellent example of how a state agency was able to scale ERHMS to a manageable size and to deploy the system quickly covering several areas: rostering, responder activities, health and safety monitoring and surveillance including data analysis and reporting, and after-action assessments. It certainly helped that they were already in the planning and development phases for such a system and that they had the information technology resources in-house to do it quickly.

Centers for Disease Control and Prevention

Before the 2014 Ebola response, the Centers for Disease Control and Prevention (CDC) had a responder health and safety program called "Responder Readiness" that addressed all three phases of deployment. In the predeployment phase, responders were medically cleared and vaccinated according to what they needed for the country they were deploying too; respirator fit testing was performed if needed; responders selected the deployment role that they wanted to deploy in, and the deployment coordinators ensured that they deployed in their selected roles. Tiered training was available based on selected deployment role as well. During the deployment phase, responders were told to report injuries or illnesses to their team lead. Exposure assessment and health monitoring and surveillance was conducted, but it was focused on the affected population, not on CDC staff. In the postdeployment phase, CDC had a robust after-action program, and responders were told that they could go to the Employee Assistance Program (EAP) if they needed help processing what happened during their deployments.

Soon after the 2014 Ebola response started, first responders came back from West Africa with serious concerns about health and safety. Five working groups were formed and made health and safety recommendations. The work groups were pre-, during, and postdeployment, mental health, and medical evacuation (medevac). One of the common themes across the five work groups was the recommendation to implement ERHMS more fully at CDC. The Deployment Risk Mitigation Unit (DRMU) was created to address these recommendations and is analogous to the ERHMS Unit described in the ERHMS document. Pre-, during, and postdeployment coordinator positions were created to oversee these three areas of deployment. Safety officers were deployed to the three affected countries to ensure the health and safety of responders. Health communicators were brought in to help address stigma issues faced on returning home. For example, for some, their child's day care would not let their child go to day care until the responder had completed the 21-day monitoring period; for others, their coworkers wanted them to telework during the monitoring period rather than return to the office.

In the predeployment phase, mental health screening was added. A thorough 4-h predeployment briefing on health and safety was required for all prior to deployment. Also, there was an improved staffing process where potential responders were asked to submit their resumes, and these were used to match responders to deployment roles.

During deployment, health and safety monitoring and surveillance was primarily conducted by in-country safety officers. Their responsibilities included: personnel safety and security in consultation with the Regional Security Officer at each Embassy, risk assessments and request and coordination for medevac if needed, injury and illness surveillance, food and water safety, vector and rodent control, lodging assessments, resiliency, and accountability. The safety officers played a critical role in ensuring the health and safety of responders, at times with over 100 responders in country, and responded to potential issues quickly.

Once responders returned to CDC, they attended debriefs covering resiliency and the EAP, after-action, and security. These debriefs were crucial since it was difficult to communicate with responders while they were in West Africa. Responders might also attend a separate role-specific debriefing. CDC coordinated active fever monitoring for all deployed staff and reported the information to all states and territories as needed.

This example demonstrates how CDC assessed their responder health and safety program and used ERHMS to identify gaps and fill them for the 2014 Ebola response. The DRMU was created based on the ERHMS Unit described in the ERHMS document and may serve as a model for others to use, particularly the creation of the pre-, during, and postdeployment coordinator positions. It also highlights the importance of safety officers deploying to the field and postdeployment debriefs. Most importantly, these improvements have been made a permanent part of CDC's responder health and safety program going forward.

SUMMARY

In past incidents, emergency responders' health and safety has not been protected as well as it should. The ERHMS framework provides guidance and recommendations about how to effectively protect them during all phases of deployment: pre-, during, and postdeployment. ERHMS is being successfully applied at both state and federal response agencies. Its use at GA DPH during the Hurricane Matthew Response provides an example of how ERHMS can be scaled to a manageable level and implemented quickly to protect public health responders. During the 2014 Ebola response, CDC improved its existing responder health and safety program by assessing the existing system, identifying gaps and filling them using the ERHMS framework.

References

Agency for Toxic Substances and Disease Registry. (2006). *Rapid response registry*. Available at https://www.atsdr.cdc.gov/rapidresponse/index.html.

Bergan, T., Thomas, D., Schwartz, E., McKibben, J., & Rusiecki, J. (December 2015). Sleep deprivation and adverse health effects in United States coast guard responders to hurricane Katrina and Rita. *Sleep Health*, *1*(4), 268–274.

Centers for Disease Control and Prevention. (April 28, 2006). Health hazard evaluation of police officers and firefighters after Hurricane Katrina—New Orleans, Louisiana, October 17–28 and November 30–December 5, 2005. *Morbidity and Mortality Weekly Report, 55*(16), 456–458.

Centers for Disease Control and Prevention. (2017). *Epi-info*. Available at https://www.cdc.gov/epiinfo/index.html.

Centers for Disease Control and Prevention. (2011). *Public health preparedness capabilities: National standards for state and local planning, March 2011*. Available at https://www.cdc.gov/phpr/capabilities/DSLR_capabilities_July.pdf.

Georgia Department of Public Health. (2017). *Disease reporting*. Available at https://dph.georgia.gov/disease-reporting.

Herbert, R., Moline, J., Skloot, G., Metzger, K., Baron, S., Luft, B., et al. (2006). The World Trade Center disaster and the health of workers: Five-year assessment of a unique medical screening program. *Environmental Health Perspectives, 114*(12), 1853–1858.

Kitt, M. M., Decker, J. A., Delaney, L., Funk, R., Halpin, J., Tepper, A., et al. (July 2011). Protecting workers in large-scale emergency responses: NIOSH experience in the Deepwater Horizon response. *Journal of Occupational and Environmental Medicine, 53*(7), 711–715.

Lowe, S. R., Kwok, R. K., Payne, J., Engel, L. S., Galea, S., & Sandler, D. P. (2015). Mental health service use by cleanup workers in the aftermath of the Deepwater Horizon oil spill. *Social Science & Medicine, 130*, 125–134.

National Institute for Occupational Safety and Health (NIOSH). (2004). Protecting emergency responders. In *Safety management in disaster and terrorism response* (Vol. 3). Cincinnati, OH: U.S. Department of Health and Human Services, Centers for Disease Control and Prevention, National Institute for Occupational Safety and Health. DHHS (NIOSH) Publication No. 2004-144. RAND Publication No. MG-170 http://www.cdc.gov/niosh/npptl/guidancedocs/rand. html.

National Response Team. (2012a). *Emergency responder health monitoring and surveillance: National Response team Technical Assistance Document (TAD), 01/26/12*. Available at http://ERHMS.nrt.org.

National Response Team. (2012b). *Emergency responder health and surveillance: A guide for key decision makers, 1/26/12*. Available at http://ERHMS.nrt.org.

Newman, D. M. (2014). Protecting worker and public health during responses to catastrophic disasters—learning from the World Trade Center experience. *American Journal of Industrial Medicine, 57*, 1285–1298.

Reissman, D. B., Funk, R., Halpin, J., Piacentino, J., Kitt, M. M., & Howard, J. (2012). Chapter 19: Protecting emergency response and recovery workers. In B. S. Levy, & V. W. Sidel (Eds.), *Terrorism and public health: A balanced approach to strengthening systems and protecting people* (2nd ed.). Oxford Press.

Rusiecki, J. A., Thomas, D. L., Chen, L., Funk, R., McKibben, J., & Dayton, M. R. (August 2014). Disaster-related exposures and health effects among US Coast Guard responders to Hurricanes Katrina and Rita: A cross-sectional study. *Journal of Occupational and Environmental Medicine, 56*(8), 820–833.

US Department of Health and Human Services. (2005). *Interim technical and policy guidelines, standards, and definitions: System development tools for the emergency system for Advance Registration of Volunteer Health Professionals (ESAR-VHP) program*. Available at http://www.kdheks.gov/it_systems/K-SERV/ESAR-VHP_Guidelines.pdf.

Further Reading

Moline, J. M., Herbert, R., Levin, S., Stein, D., Luft, B. J., Udasin, I. G., et al. (2008). WTC medical monitoring and treatment program: Comprehensive health care response in aftermath of disaster. *Mount Sinai Journal of Medicine, 75*, 67–75.

Vignette: Experiences Working With Ebola Virus Disease and Pregnancy in Sierra Leone, 2014

Jonetta Johnson Mpofu[1,2], *Michelle Dynes*[1,2]

[1]Centers for Disease Control and Prevention, Atlanta, GA, United States; [2]US Public Health Service Commissioned Corps, Rockville, MD, United States

During the 2014 Ebola virus disease (EVD) epidemic, teams of health workers, scientists, laboratory technicians, and local and international volunteers worked to decrease EVD transmission and meet the needs of people suffering from and at risk for EVD. As members of the Centers for Disease Control and Prevention's (CDC) international response team and epidemiologists in the Division of Reproductive Health, we evaluated data to identify opportunities to improve care of pregnant women who were suffering from or at risk for EVD. Our work took place from August to December of 2014 in Sierra Leone.

Sierra Leone has the highest maternal mortality ratio in the world, with an estimated 1360 maternal deaths per 100,000 live births in 2015 (World Health Organization, 2015b). Prior to the 2014 EVD epidemic, Sierra Leone was making great strides to improve maternal and infant health. In 2010, Sierra Leone established a free health-care program, the Free Health Care Policy, for pregnant and lactating women to increase access to antenatal care, facility-based deliveries, and postnatal care (Amnesty International, 2011; UNFPA, 2015). Results of the policy were encouraging as antenatal care visits, postnatal care visits, and the number of deliveries in health facilities increased (Amnesty International, 2011; UNFPA, 2015).

The arrival of EVD disrupted progress in improving maternal and infant health outcomes in Sierra Leone by contributing to increases in maternal and infant death, causing health worker shortages, and reducing the number of women receiving antenatal care and delivering services in health facilities (Assessment Capacities Project (ACAPS) (2015); Dynes, Miller, Sam, Vandi, & Tomczyk, 2015; UNFPA, 2015; World Health Organization, UNICEF, UNFPA, The World Bank, & UN Populations Division, 2014). In Sierra Leone, an estimated 69% of all health-care workers with confirmed or probable EVD died

Disaster Epidemiology
http://dx.doi.org/10.1016/B978-0-12-809318-4.00017-4

(World Health Organization, 2015a)— an impact that is estimated to potentially increase the maternal mortality ratio by 74% as compared with pre-EVD (Evans, Goldstein, & Popova, 2015). Below we describe data collection activities aimed at improving screening for EVD and care for pregnant women during the 2014 EVD epidemic in Sierra Leone.

ACTIVITY 1. IMPROVING SCREENING CRITERIA FOR EBOLA VIRUS DISEASE AMONG PREGNANT WOMEN

Pregnant women infected with EVD are at increased risk of severe illness, pregnancy complications, poor perinatal outcomes, and death (Black, Caluwaerts, & Achar, 2015; Caluwaerts et al., 2014; Mupapa et al., 1999). Identification of EVD in pregnant women may be difficult due to an overlap in symptoms of pregnancy/labor and EVD. Efforts to reduce excess morbidity and mortality associated with EVD among pregnant women during outbreaks are especially crucial in settings with high maternal mortality and struggling health systems pre-outbreak (Hayden, 2015; Statistics Sierra Leone (SSL) & ICF International, 2014; UNFPA, 2015; World Health Organization, UNICEF, UNFPA, The World Bank, & UN Populations Division, 2014).

EVD screening criteria for the general population were well established before the 2014 EVD epidemic. However, a screening protocol for pregnant women had not been developed and information on the clinical presentation and symptomatology was critically needed to improve screening, diagnosis, and prognosis and reduce morbidity and mortality of pregnant women with suspected EVD.

To address this need, we collected data on symptoms of EVD in pregnant women admitted to isolation units in Sierra Leone and examined associations between EVD symptoms and EVD status. Our team of CDC staff and local health workers visited seven EVD isolation units across Sierra Leone from July–December, 2014. Women were admitted to EVD isolation units if they screened positive for ≥1 of the following criteria: contact with a family member with EVD, lived in a quarantined home, attended a funeral of someone with EVD, were a health worker, or had a current fever (temperature >38°C) and ≥3 of the following self-reported symptoms suggestive of EVD: headache, nausea/vomiting, difficulty breathing/swallowing, fatigue/weakness, muscle/joint pain, diarrhea, abdominal pain, anorexia and bleeding. We abstracted data from medical records of 192 pregnant women suspected of having EVD. EVD status was determined by polymerase chain reaction testing and reported as positive or negative after women were admitted to isolation units.

We examined associations between EVD status and symptoms using chi-square tests and multivariate logistic regression. Preliminary results showed most women were EVD negative and either not in labor or had no documentation about their labor status. Thirty-six percent of women were EVD positive. The top three reported EVD symptoms recorded at admission among all women were abdominal pain, fever and fatigue/weakness(Johnson et al., 2015). Several symptoms were more common in EVD positive than negative women (Johnson et al., 2015).

Preliminary conclusions note high variability in symptoms by EVD status. Such variability may be due to study limitations such as retrospective data collection, which may be subject to recall bias, and missing information, which resulted in data that were not discriminant enough to guide clinical decisions. There is a continued need for additional data on the clinical presentation of pregnant women during EVD outbreaks to improve and refine EVD screening of pregnant women. Pregnant women and their fetuses represent a vulnerable population group with unique health-care needs and increased health risks during disaster and emergency situations (Hoffman, 2009; World Health Organization, UNICEF, UNFPA, The World Bank, & UN Populations Division, 2014). It is important to develop group specific screening criteria and care protocols for Vulnerable Groups (World Health Organization, 2017), such as pregnant women, during disaster and emergency situations. Better understanding of the symptomatology and progression of EVD in pregnant women prior to the 2014 EVD epidemic may have helped to reduce mortality and morbidity for pregnant women, their fetuses, and respective health care workers.

ACTIVITY 2. ADDRESSING BARRIERS TO CARE AMONG PREGNANT WOMEN DURING AN EVD OUTBREAK

Kenema, located in the Eastern Province of Sierra Leone, was one of the earliest and most heavily burdened districts during the 2014 Ebola epidemic. From May to July, 2014, one-third of all confirmed Ebola cases across Sierra Leone originated in Kenema (Sierra Leone Ministry of Health and Sanitation (MOHS), unpublished data, 2014). An Ebola Treatment Unit (ETU) was established at Kenema General Hospital (KGH), and in the early months of the epidemic, Ebola patients from across the country were transported to KGH for testing and treatment.

Health services data demonstrated that routine maternal and newborn health service use had declined in Kenema between May and June 2014 (Sierra Leone MOHS, unpublished data, 2014). In September 2014, the Kenema District Health Management Team, International Rescue Committee (IRC), and the CDC collaborated on a project with two objectives: (1) to better understand factors that might have contributed to these declines; and (2) to explore approaches to increase use of routine services. We carried out nine focus group discussions with 34 health workers and 27 pregnant and lactating women at six primary health-care facilities in Kenema District. We used a structured interview guide to ask open-ended questions related to health facility use, reasons for decreased facility use, ideas for encouraging women to return to care, and perceptions of safety. We then performed a rapid content analysis on interview notes to group responses into common themes.

Health workers and clients reported a sharp decline in facility use for routine health services immediately after the epidemic began. All participants believed that the main reason for decreased use of facilities was fear of contracting Ebola. Common misconceptions early in the epidemic contributed to these fears, such as the erroneous belief that staff was paid for each client referred to the ETU, and thus everyone who went to a health facility would be taken to the Kenema ETU. Ebola awareness and educational activities were implemented to increase health care seeking behavior. Health workers and clients believed that recent infection prevention and control trainings, along with equipment for providers, increased use of services by providing reassurance to the community. Community-based approaches to encourage women and children to seek health services were popular among participants. One woman recommended having clients who had recently received care in the facility return to their villages to share their positive experiences. Traditional birth attendants showed a deep interest in helping to share Ebola messaging by going into the center of villages and singing and dancing as a way to call women together for education.

Information from this project highlighted the need to restore community member's confidence in their health facilities. Engaging the community after infection prevention and control trainings became a key strategy to encourage women and families to return to health facilities (Dynes et al., 2015). One of our team's key lessons learned from this experience was that interorganization collaboration and coordination are absolutely critical for success. For example, the health ministry partners provided logistical support, while the locally based IRC partners drew on their community connections to help facilitate and translate group discussions, and CDC analyzed interview notes and drafted the report. Looking back at this critical time in Kenema District, we emphasize the importance of being able to implement rapid data collection, analysis, and development of recommendations in the ever changing environment of an emergency. Quality information for decision-making is a critical step that cannot be overlooked even when responding to public health crises.

Acknowledgments

Dr. Fatma Soud, Dr. Meghan Lyman, Dr. Alimamy P. Koroma, Dr. Diane Morof, Sascha Ellington, Dr. Samuel S. Kargbo, Dr. William Callaghan, Laura Miller, Tamba Sam, Dr. Mohammed Alex Vandi, and Dr. Barbara Tomczyk.

Disclaimer

The findings and conclusions in this report are those of the authors and do not necessarily represent the official position of the Centers for Disease Control and Prevention.

References

Amnesty International. (2011). *At a crossroads: Sierra Leone's free health care policy*. London: Amnesty International Ltd. Retrieved from https://www.amnestyusa.org/files/pdfs/sierral_maternaltrpt_0.pdf.

Assessment Capacities Project (ACAPS). (2015). *Thematic note—ebola outbreak in West Africa: Impact on health service utilization in Sierra Leone*. Thematic Note. ACAPS. Retrieved from https://www.acaps.org/sites/acaps/files/products/files/f_impact_on_health_service_utilisation_in_sierra_leone_march_2015.pdf.

Black, B. O., Caluwaerts, S., & Achar, J. (2015). Ebola viral disease and pregnancy. *Obstetric Medicine, 8*(3), 108—113.

Caluwaerts, S., Lagrou, D., Van Herp, M., Black, B., Caluwaerts, A., Taybi, A., et al. (2014). *Guidance paper Ebola Treatment Center (ETC): Pregnant and lactacting women*. Brussels: MSF. Retrieved from https://www.rcog.org.uk/globalassets/documents/news/etc-preg-guidance-paper.pdf.

Dynes, M. M., Miller, L., Sam, T., Vandi, M. A., & Tomczyk, B. (2015). *Perceptions of the risk for ebola and health facility use among health workers and pregnant and lactating women — Kenema district, Sierra Leone, September 2014*. Atlanta, GA. Retrieved from https://www.cdc.gov/mmWr/preview/mmwrhtml/mm6351a3.htm.

Evans, D. K., Goldstein, M., & Popova, A. (2015). Health-case worker mortality and the legacy of the ebola epidemic. *Lancet Glob Health, 3*(8), e439—440.

Hayden, E. C. (2015). Ebola's lasting legacy. *Nature, 519*, 24—26.

Hoffman, S. (2009). *Preparing for disaster: Protecting the most vulnerable in emergencies*. Retrieved from https://lawreview.law.ucdavis.edu/issues/42/5/articles/42-5_Hoffman.pdf.

Johnson, J., Callaghan, W., Hayfa, E., Ellington, S., Kargbo, S. S., Koroma, A. P., et al. (2015). Clinical presentation of pregnant women at isolation centers for ebola virus disease in Sierra Leone, 2014. In *Paper presented at the American public health association, Chicago, IL*. https://apha.confex.com/apha/143am/webprogram/Paper338080.html.

Mupapa, K., Mukundu, W., Bwaka, M. A., Kipasa, M., De Roo, A., Kuvula, K., et al. (1999). Ebola hemorrhagic fever and pregnancy. *Journal of Infectious Disease, 179*(Suppl. 1), 22–23.

Statistics Sierra Leone (SSL), & ICF International. (2014). *Sierra Leone demographic and health survey, 2013*. Freetown, Sierra Leone and Rockville, Maryland, USA. Retrieved from https://dhsprogram.com/pubs/pdf/FR297/FR297.pdf.

UNFPA. (2015). *Rapid assessment of ebola impact on reproductive health services and service seeking behavior in Sierra Leone*. Freetown, Sierra Leone. Retrieved from http://www.mamaye.org.sl/sites/default/files/evidence/UNFPA%20study%20_synthesis_March%2025_final_d.pdf.

World Health Organization. (2015a). *Health worker ebola infections in Guinea, Liberia and Sierra Leone*. Geneva, Switzerland. Retrieved from http://apps.who.int/iris/bitstream/10665/171823/1/WHO_EVD_SDS_REPORT_2015.1_eng.pdf?ua=1&ua=1.

World Health Organization. (2015b). *Trends in maternal mortality: 1990 to 2015: Estimates by WHO, UNICEF, UNFPA, World Bank Group and the United National Population Division*. Geneva, Switzerland. Retrieved from http://apps.who.int/iris/bitstream/10665/194254/1/9789241565141_eng.pdf?ua=1.

World Health Organization, UNICEF, UNFPA, The World Bank, & UN Populations Division. (2014). *Trends in maternal mortality: 1990–2013. Estimates by WHO, UNICEF, UNFPA, the World Bank, and the United Nations Population Division*. Geneva. Retrieved from http://apps.who.int/iris/bitstream/10665/112682/2/9789241507226_eng.pdf.

World Health Organization. (2017). *Vulnerable Groups. Environmental Health in Emergencies*. Retrieved from http://www.who.int/environmental_health_emergencies/vulnerable_groups/en/.

CHAPTER

12

Methods: Data Analysis for Disaster Epidemiology

Ashley Conley

St. Joseph Hospital, Nashua, NH, United States

DATA ANALYSIS FOR DISASTER EPIDEMIOLOGY

Data that are collected and analyzed in the preparedness, response, recovery, and mitigation phases of the disaster management cycle can be a vital source of information for a community. Analysis of disaster epidemiology data can answer many questions commonly posed by planners and emergency managers such as:

- How resilient is my community to respond to a disaster?
- Is there a need for injury prevention measures posthurricane?
- What factors put my residents more at risk from a tornado?
- What actions did residents take that kept them safe during the disaster?

Various data collection tools such as surveillance systems, surveys, assessments, and epidemiologic studies can provide the data needed to answer these questions and many more. The data collected can provide vital information to planners, decision-makers, and emergency managers and can assist in making informed decisions and potentially save lives. To analyze the data that are collected before, during, and after a disaster, descriptive epidemiology and analytic epidemiology are used to answer the what, who, where, when, how, and why diseases, injuries, and exposures are increasing or decreasing (USDHHS, 2012). In this chapter, we will review how these tools are used to analyze data.

Descriptive Epidemiology: Data Analysis

Data analysis for descriptive epidemiology includes using counts, proportions, and rates to determine the person, place, and time that could be affected before a disaster or are being affected during a disaster (USDHHS, 2012). Counts consist of a tally of events, cases, or resources that can assist emergency managers with requesting supplies to distribute to an

Disaster Epidemiology
http://dx.doi.org/10.1016/B978-0-12-809318-4.00018-6

affected community or recognizing when there is an above normal increase in events, such as an increase in the number of emergency department visits following a disaster.

Proportions are expressed as a decimal, percentage, or a fraction where the numerator is also included in the denominator. They are used to identify the quantity of people affected or amount of resources utilized (USDHHS, 2012). For example, if you have 100 volunteers readily available to deploy to an emergency shelter and you have already deployed 25 of them. You have utilized 0.25 (decimal) or 1/4 (fraction) or 25% (percentage) of your volunteer pool for the emergency shelter. In the same way, a community experiencing a power outage could have 80% of its households without power and 20% with power.

Following Hurricane Ike in 2008 in Texas, the CDC utilized a variety of surveillance systems to determine carbon monoxide exposures, which utilized descriptive epidemiology, counts, and percentages in the data analysis. In this disaster, the National Poison Data System and the Texas Poison Center Network identified 54 calls to the centers that described storm-related carbon monoxide exposures following the hurricane. Hampson et al. (2009) described the demographics of the callers by age (median 24 years), gender (64.8% female), and types of symptoms such as nausea (44%) and vomiting (28%). In these cases, 82% identified a generator as a source of carbon monoxide exposure and most (93%) were in a residential setting (Hampson et al., 2009). This study identified trends and commonalities using existing surveillance systems to inform education efforts in future power outages and is an example of the use of descriptive epidemiology to identify person, place, and time.

Although counts and proportions are helpful, rates can provide a comparison based on population size that cannot be accomplished with counts and proportions. Rates are "measures that relate the number of cases during a certain period of time to the size of the population in which they occurred" (CDC, 2016a; USDHHS, 2012). Rates can be compared across different geographies, populations, and time periods. Rates can be calculated for incidence, prevalence, mortality, and morbidity (CDC, 2016a). Incidence refers to the number of new cases of an illness, injury, death, or condition in a population over a period of time and prevalence refers to the number that already existed during a period of time (CDC, 2016a; USDHHS, 2012).

Incidence proportion, also referred to as the attack rate, is the proportion of the population that develops an illness during an outbreak and can be used following disasters when there are outbreaks in the affected communities. The numerator includes the number of new cases of a disease and the denominator includes the size of the population. In foodborne outbreaks, a food-specific attack rate can be calculated where the numerator includes the number of people who ate a specific food and became ill and the denominator includes the number of people who ate the specific food (USDHHS, 2012). Emergency shelters that are set up to house evacuees following disasters can become prime settings for outbreaks and foodborne illness if appropriate prevention measures are not implemented and maintained. Following Hurricane Katrina in 2005, a norovirus outbreak occurred among evacuees at the Reliant Park emergency shelter in Texas where 1169 evacuees, about 18% of evacuees that visited the shelter, reported symptoms of diarrhea and vomiting (Palacio et al., 2005).

When health data are available predisaster, for example, the prevalence rate of physical and mental health conditions by gender, age, and location, the rates postdisaster can be calculated and compared with predisaster data to look for trends and changes in the community. In a study by Swerdel, Janevic, Cosgrove, Kostis, and Myocardial Infarction Data Acquisition

System (MIDAS 24) Study Group, 2014 the incidence and 30-day mortality were calculated for cardiovascular events, including myocardial infarction and stroke following Hurricane Sandy in New Jersey. The data demonstrated that after Hurricane Sandy, there was a 22% increase in incidence of myocardial infarctions, a 31% increase in the 30-day mortality from myocardial infarction and a 7% increase in the incidence stroke, and no change in the 30-day mortality from stroke (Swerdel et al., 2014).

Analysis of Survey Data

Surveys are conducted in a multitude of ways before, during, and after a disaster to collect information. They are collected via phone, face-to-face, sent in the mail, or posted online. An example of a survey that is conducted yearly by the CDC, regardless of disasters, is the Behavioral Risk Factor Surveillance System (BRFSS) which is a state-based, random-digit-dialed telephone survey of the noninstitutionalized US civilian population ≥18 years (DeBastiani, 2012). DeBastiani, et al., looked at BRFSS data collected from 14 states from 2006 to 2010 that asked residents about their household preparedness levels. The study showed that 94.8% of households had a working battery-operated flashlight but only 21% had a written evacuation plan (DeBastiani, 2012). The data collected in this survey can be utilized by planners and emergency managers to educate their communities about preparedness and encourage them to have disaster kits and plans in place to prepare for a disaster.

Analysis of Data for the Assessment of Chemical Exposures Program

One program that utilizes surveys as a major component of the data gathering process is the Assessment of Chemical Exposures (ACE) Program, which conducts assessments using surveys and medical chart abstraction to understand who was exposed, where they were exposed, and what health affects the exposed individuals experienced as a result of their exposure. To gather data, face-to-face interviews can be conducted or data can be collected using paper-based or mobile technology. Surveys can also be distributed for individuals to complete. Questions that ask about the types of symptoms, severity of symptoms, and access to health care can help provide the descriptive information needed to understand the scenario and create recommendations to prevent future incidents (CDC, 2015). In an event at a metal recycling facility in California in 2010, the data collected were stratified by gender, age, ethnicity, education, symptoms, preexisting health conditions, and location at the time of the event and presented as percentages. In brief, the report stated that 93% (27 out of 29) of the exposed persons were >18 years, 56% were outdoors at the time of the chlorine gas release, and 19% had preexisting high blood pressure (Kelsey et al., 2011).

Analysis of Data for the Community Assessment for Public Health Emergency Response

The Community Assessment for Public Health Emergency Response (CASPER) is a household-based survey that uses a two-stage cluster sample to obtain 210 surveys from 30 clusters in a community. Data collection for a CASPER can be done using paper and pen or mobile technology. Once uploaded into an electronic database, the data collected can be manipulated and analyzed, and visuals such as graphs can be created to help various audiences understand the data. Prior to analysis, the data should undergo a quality assurance

TABLE 12.1 Data Analysis From Community Assessment for Public Health Emergency Response (CASPER) Following Microcystin Toxin Contamination in Ohio (McCarty et al., 2016)

Characteristic	No.	Estimated No. of Households	% of Sampled Households (95% CI)
HOUSEHOLD PHYSICAL HEALTH SYMPTOMS			
Any illness	25	17,431	16.2 (7.6–24.8)
Nausea	16	9,833	9.1 (4.2–14.0)
Vomiting	10	6,739	6.2 (1.9–10.5)
Abdominal pain	11	8,028	7.4 (2.5–12.4)
HOUSEHOLD MENTAL HEALTH SYMPTOMS			
Any mental health concerns	14	10,443	9.9 (4.4–15.4)
Anxiety or stress	10	7,607	7.0 (2.3–11.8)
Loss of appetite	5	5,243	4.8 (0.2–9.5)

process to ensure the information entered into the database is correct (CDC, 2012). Additionally, to reduce bias, a weight variable based on the probability of being selected is added to each surveyed household. Weighted estimates with 95% confidence intervals and unweighted estimates can be calculated for each measure. Table 12.1 highlights data analysis from a CASPER that was completed by the Ohio Department of Health and the Toledo-Lucas County Health Department following the increase of microcystin toxin levels in the drinking water. The symptoms experienced by household members, such as nausea and vomiting, and mental health effects, such as anxiety and loss of appetite, were included in the survey and analysis. About 16% of households experienced some type of illness and about 10% experienced mental health concerns (McCarty et al., 2016).

The CDC's CASPER toolkit includes three calculations for response rates: completion rate, cooperation rate, and contact rate (Table 12.2). The completion rate is used to determine what percent of surveys that were expected to be completed were actually completed. The goal is to reach 100% of the 210 surveys expected to be completed in the CASPER protocol. The

TABLE 12.2 Response Rate Calculations for Community Assessment for Public Health Emergency Response (CASPER)

Completion rate	$\frac{\text{Number of completed interviews}}{\text{Number of interviews (usually 210)}}$
Cooperation rate	$\frac{\text{Number of completed interviews}}{\text{All housing units where contact was made}}$ Contact includes completed interviews, incomplete interviews, and refusals
Contact rate	$\frac{\text{Number of completed interviews}}{\text{Number of housing units where contact was attempted}}$ Attempted contact includes completed interviews, incomplete interviews, refusals, and nonrespondents

cooperation rate is the proportion of households at which contact was made and the household agreed to complete an interview and the contact rate is the proportion of all households at which contact was attempted and the household successfully completed an interview (CDC, 2012).

Continuing the example from Ohio, the teams completed 171 surveys for an 81% completion rate and contacted 314 households for a 54.5% cooperation rate (McCarty et al., 2016).

Analytic Epidemiology: Data Analysis

Descriptive epidemiology and the methods for data analysis described so far provide credible, reliable data that can be used by decision-makers to identify trends and make decisions. From these observations, hypotheses can be developed and tested to better understand the risk factors and causality of the exposure to certain health outcomes. Studies used in analytic epidemiology test the hypotheses and focus on the why and how diseases and conditions affect a population (USDHHS, 2012). We will review two types of observational studies that can be conducted pre- and postdisaster, case-control studies and cohort studies.

Cohort Studies

Cohort studies work well when you have a defined group of people that were exposed or not exposed who can be reached to be a part of the study. The participants are then studied over time to monitor for disease occurrence, in a prospective study, or are asked about their history in a retrospective study. In both scenarios, the goal is to compare the rates of disease in the exposed group and the unexposed group to better understand risk factors and the relationship to health outcomes (USDHHS, 2012). The measure of association used in cohort studies is a risk ratio, also known as relative risk. The risk ratio is calculated by dividing the risk (incidence proportion) in the exposed group and the risk in the unexposed group (Eq. 12.1). If the risk ratio is 1.0, there is no difference in the risk between the exposed and unexposed groups. If it is greater than 1.0 then there is an increased risk for the exposed group and if it is less than 1.0 then there is a reduced risk for the exposed group (USDHHS, 2012).

A two-by-two table is often used to orient the data in an epidemiologic study. An example of a two-by-two table is shown below along with the calculation (Eqs. 12.1) (USDHHS, 2012).

$$\text{Risk Ratio} = \frac{\text{Incidence in exposed}}{\text{Incidence in unexposed}} = \frac{\frac{A}{A+B}}{\frac{C}{C+D}} \tag{12.1}$$

Following disasters, prospective cohort studies can be completed when researchers are trying to understand if there are any long-term health effects from the exposure. Unexposed individuals are used as a comparison group. A well-known and long-term example of a cohort study following a disaster is the World Trade Center (WTC) Health Registry, which was developed following the 9/11 terrorist attacks in New York City at the WTC. This is the largest registry developed and maintained following a disaster in US history, with the objective to study the long-term health effects of living and working near the WTC following the attacks (NYC 9/11 Health, 2016).

Case-Control Studies

Case-control studies are commonly used when the route of exposure is unknown and there is not a clearly defined group of people that can be identified as exposed and unexposed. In these studies, cases are individuals that have the disease and controls are those that do not have the disease. The measure of association for case-control studies is the odds ratio. The data to calculate an odds ratio can be put in a two-by-two table, as shown in Table 12.3. The calculation for the odds ratio is shown below (Eq. 12.2) (USDHHS, 2012).

$$\text{Odds Ratio} = \frac{\text{Odds that the cases were exposed}}{\text{Odds that the controls were exposed}} = \frac{ad}{bc} \tag{12.2}$$

In a case-control study conducted by Ward, Spokes, and McAnulty (2011), the authors looked at risk factors for hospitalization in Sydney, Australia, due to the 2009 H1N1 influenza A pandemic strain in individuals >16 years old (Ward et al., 2011). The pandemic strain of H1N1 emerged in 2009 and spread quickly across the globe, disproportionately affecting children and young adults. During the 2009–10 influenza season, there was an increase in pediatric deaths and hospitalizations for children and young adults not traditionally seen in prior influenza seasons (World Health Organization Collaborating Center for Surveillance, Epidemiology, and Control of Influenza, 2010). The case-control study analyzed data from 302 case-patients and 603 controls and identified pregnancy, immune suppression, preexisting lung disease, asthma, heart disease, diabetes, and smoking as risk factors for hospitalization from July 1, 2009–August 31, 2009 (Ward, Spokes, & McAnulty, 2011).

Epi Info

To quickly and efficiently analyze data, epidemiologists need access to a statistical software package and a database to store the data that are collected. CDC's Epi Info is free and can be used by epidemiologists and public health organizations to analyze data. Epi Info is a downloadable resource that can be used on computers, mobile devices, and the Web. Data entry forms can be customized and developed in the software, which can easily analyze data and be used to create graphs and maps. It also has an interactive dashboard that can be used to visualize data and be a resource for emergency managers (CDC, 2016b). It is a resource with many benefits and additional information about the product can be found on the CDC website at http://www.cdc.gov/epiinfo/index.html.

TABLE 12.3 Example Two-by-Two Table

	Ill	Not Ill	Total
Exposed	A	B	A + B
Unexposed	C	D	C + D
Total	A + C	B + D	

CONCLUSION

Public health emergency preparedness professionals, emergency managers, and other disaster response agencies should consider how data collection and analysis can assist them through the disaster management cycle to maintain situational awareness of the health status of the population. The data collection and analysis process that is used should be tailored to meet the needs of the disaster scenario and modified as needed. Used efficiently and correctly, data analysis has the potential to save lives and leverage scarce resources.

References

CDC. (2015). *Assessment of Chemical Exposures (ACE) program*. Retrieved from http://www.atsdr.cdc.gov/ntsip/ace.html.

CDC. (2016a). *A primer for understanding the principles and practices of disaster surveillance in the United States* (1st ed.). Atlanta (GA): CDC.

CDC. (2016b). *Epi Info*. Retrieved fromhttp://www.cdc.gov/epiinfo/index.html.

Centers for Disease Control, Preparedness (CDC). (2012). *Community Assessment for Public Health Emergency Response (CASPER) toolkit* (2nd ed). Atlanta (GA): CDC.

DeBastiani, S. D. (2012). Household preparedness for public health emergencies – 14 States, 2006–2010. *Morbidity and Mortality Weekly Report, 61*(36), 713–719.

Hampson, N., Dunn, S., Bronstein, A., Fife, C., Villanacci, J., Zane, D., et al. (2009). Carbon monoxide exposures after hurricane Ike-Texas, September 2008. *Morbidity Mortality Weekly Report, 85*(31), 845–849.

Kelsey, K., Roisman, R., Kreutzer, R., Materna, B., Duncan, M., Orr, M., et al. (2011). Chlorine gas exposure at a metal recycling facility – California, 2010. *Morbidity and Mortality Weekly Report, 60*(28), 951–954.

McCarty, C., Nelson, L., Eitniear, S., Zgodzinski, E., Zabala, A., Billing, L., et al. (2016). community needs assessment after microcystin toxin contamination of a municipal water supply — Lucas County, Ohio, September 2014. *Morbidity Mortality Weekly Report, 65*, 925–929. http://dx.doi.org/10.15585/mmwr.mm6535a1.

NYC 9/11 Health. (2016). *World Trade Center health registry*. Retrieved fromhttp://www1.nyc.gov/site/911health/about/wtc-health-registry.page.

Palacio, H., Shah, U., Kilborn, C., Martinez, D., Page, V., Gavagan, T., et al. (2005). Norovirus outbreak among evacuees from Hurricane Katrina – Houston, Texas, September 2005. *Morbidity and Mortality Weekly Report, 54*(40), 1016–1018.

Swerdel, J. N., Janevic, T. M., Cosgrove, N. M., Kostis, J. B., & Myocardial Infarction Data Acquisition System (MIDAS 24) Study Group. (2014). The effect of Hurricane Sandy on cardiovascular events in New Jersey. *Journal of the American Heart Association, 3*(6), e001354.

U.S. Department of Health and Human Services (US DHHS). (2012). *Principles of epidemiology in public health practice* (3rd ed.). Atlanta (GA): CDC.

Ward, K. A., Spokes, P. J., & McAnulty, J. M. (2011). Case-control study of risk factors for hospitalization caused by pandemic (H1N1) 2009. *Emerging Infectious Diseases, 17*(8), 1409–1416.

World Health Organization Collaborating Center for Surveillance, Epidemiology, and Control of Influenza, et al. (2010). Update: Influenza activity – United States, 2009–10 season. *Morbidity and Mortality Weekly Report, 59*(29), 901–908.

CHAPTER

13

Applications: Biosurveillance, Biodefense, and Biotechnology

Koya C. Allen

US Department of Defense, Stuttgart, Germany

PREVENTION AND PREPAREDNESS—CONCEPTS IN BIOSURVEILLANCE

Disasters can be unpredictable; however, the consequences of disasters can be anticipated. Using knowledge about geographic location, disaster risk, and public health systems and infrastructure, epidemiologists can incorporate risk-mitigating strategies into disaster preparedness and response plans to address the public health needs of a community. Past experiences can provide lessons learned to facilitate improvements and inform frameworks that will achieve public health resiliency for future disease threats. The key is in evaluating and understanding the current public health infrastructure and regional health systems, so that new tools and interventions are leveraged to enhance early detection capabilities, support event monitoring, and improve rapid response and management of health consequences after a disaster.

What is Biosurveillance?

Biosurveillance is:

> "the process of gathering, integrating, interpreting, and communicating essential information related to all-hazards threats or disease activity affecting human, animal, or plant health to achieve early detection and warning, contribute to overall situational awareness of the health aspects of an incident, and to enable better decision-making at all levels" ***The White House, 2012.***

The National Biosurveillance Strategy defines four core functions of the biosurveillance process: (1) scan and discern the environment; (2) identify and integrate essential information; (3) inform and alert decision-makers; and (4) forecast and advise potential impacts (The White House, 2012). Biosurveillance is a dynamic process whereby the core functions are interrelated and occur simultaneously.

Disaster Epidemiology
http://dx.doi.org/10.1016/B978-0-12-809318-4.00019-8

How Does Biosurveillance Contribute to Disaster Preparedness?

Biosurveillance provides a means for anticipating a disaster as well as the consequences of that disaster or disease event. When considering the biosurveillance analytical checklist, shown in Table 13.1, each function contributes information and generates data that can inform actions to enhance preparedness. The process is not cyclical; rather, it is dynamic process that interacts to differing degrees, for continual contribution and a prolonged effect. Ensuring situational awareness is a difficult task, but it is the basis for understanding what prevention strategies exist and which ones need to be altered. In addition, as different sources of essential information are integrated, analysts can alert leadership at the appropriate levels to inform action and policy change. Forecasting should be used to project outcomes related to disaster consequences such as disease transmission risk, outbreak or epidemic potential, and impact of risk mitigation strategies (Nsoesie, Brownstein, Ramakrishnan, & Marathe, 2013). Depending on current efforts in biopreparedness or immediate needs for bioresponsiveness, the checklist can facilitate a robust biosurveillance strategy and contribute to overall disaster resiliency.

SURVEILLANCE SYSTEMS

Many will argue that there is a right and a wrong way to conduct infectious disease surveillance. Traditional surveillance programs have supported research and informed policy for generations; however, emerging technology and societal changes have created a pathway for innovation in surveillance programs. Biosurveillance is a leading innovation in developing actionable information related to infectious disease outbreaks, but the foundation of this surveillance strategy lies in the amalgamation of both event-based surveillance systems and more traditional clinical surveillance systems (Edward, Tumacha, Kerstin, Goran, & Tim, 2014). The data gathered through both mechanisms provide vital information for each core function of the biosurveillance framework and analytical checklist.

Event-based surveillance is the detection, verification, analysis, assessment, and further investigation of potential public health threats ("events"). In comparison, *indicator-based surveillance* is the systematic ongoing collection, analysis, interpretation, and dissemination of highly structured information ("indicators") for public health action (Indicator-based surveillance, n.d.). A key difference between indicator- and event-based surveillance is in the methodology and timeliness of acquiring reliable data and information (Edward et al., 2014). For event-based surveillance, timeliness is a significant benefit; the lone report or eyewitness may provide information that triggers further investigation. The quality of information, however, is often questionable as news sources and social media may not always provide the most reliable or accurate information regarding case data or outbreak dynamics. Indicator-based surveillance is often very reliable and provides data that have already gone through several layers of verification, with clearly defined case definitions and inclusion criteria. Unfortunately, timeliness is often a concern for early identification of outbreaks. Instead, these surveillance data often serve to provide detailed trend and baseline data for a population, to better understand disease dynamics in a region.

Event-based surveillance has rapidly evolved over the last 2 decades. The oldest system, Program for Monitoring of Emerging Diseases (ProMED mail) is a basic open-source Internet-based reporting system that uses media reports, official reports, local observations,

TABLE 13.1 Biosurveillance Analytical Checklist

Core Functions	Biopreparedness	Bioresponsiveness
Scan and discern the environment	• **Infectious disease baseline**—risk assessments, disease trends in humans, animals, and plants for a region • **Public health infrastructure**—Hospital capacity and capabilities; medical countermeasures available • **Current events**—new outbreaks, new technology or capabilities, security and safety risks, political climate	• **Event monitoring**—outbreak statistics including case data, contact tracing, drug supply, environmental contamination, and movement restrictions • **Rapid risk assessments**—comparison of baseline assessments with current event to determine aberrations and identify gaps in response
Identify and integrate essential information	• **Research analytics**—Use analytics to collate information, generate data, and draw conclusions • **Subjective analytics**—qualitative analysis based on expertise, experience, and logic for best practices • **Objective analytics**—quantitative scientific analysis using supportive data and measurements	• **Rapid analytics** to collate data in changing conditions • **Subjective analytics**—qualitative analysis will highlight lessons learned from past experiences in similar situations • **Objective analytics**—rapid quantitative assessments of the current situation to aid in event monitoring
Inform and alert decision-makers	• **Provide recommendations** to decision-makers at varying levels in chain of command based on essential information for preparedness activities	• **Provide recommendations** to decision-makers at varying levels in chain of command based on essential information for response activities
Forecast and advise potential impacts	• **In-depth analytics** that project forward in time to identify effective disease containment strategies, geographic susceptibility, vulnerable populations • **Generate and test preparedness plans** for effectiveness	• **Epidemic/event potential**—using infectious disease modeling and social network analysis to identify the reach of an outbreak based on current situation, intervention implementation, effectiveness, and costs

and online summaries regarding infectious diseases to disseminate vetted reports, often with commentary to further deduce the scope of an outbreak or disease situation (ProMED-mail, n.d.). Event-based surveillance programs have since evolved into more robust internet monitoring analytical systems with information scanning tools, such as the Medical Information System (MedISys). MedISys uses the Europe Media Monitor (EMM) system that has advanced information extraction techniques and collects information from news sources in over 70 languages to rapidly identify potential threats to public health using information from the Internet (Medical Information System - MEDISYS - Joint Research Centre - European Commission, n.d.) (see Table 13.2). The main limitation in all of these systems is that they are not sufficient for providing an early warning capability when used alone. Each still requires the additional component of an analyst or subject matter expert (SME), who can discern true risk using not only the information provided but also the additional epidemiological knowledge of disease pathogens and public health.

For event-based surveillance program analysis, analysts can use some or all of the various systems and tools available in their respective biosurveillance efforts. Some programs may focus on specific diseases of interest to their region or locale, and others may have interests in bioterrorism or natural emerging diseases. The key is in the application of the information generated from event-based surveillance systems for environmental situational awareness in conjunction with accurate data from indicator-based surveillance systems. Traditional surveillance systems using passive reporting of nationally notifiable diseases or active programs targeting clinical case or laboratory data on specific infectious diseases provide the baseline necessary to assist in quality aggregation of essential information. With the conclusions drawn from these systems, information can be shared with biosurveillance partners, reports can be generated with recommendations to leadership, and interventions can be developed to target disease outbreaks (Fig. 13.1). In a disaster or outbreak situation, information collection, analysis, and dissemination will have to occur rapidly. Indicator-based surveillance may increase in importance for details of disease transmission such as changes in disease incidence, isolation of a high-risk area, and effectiveness of an intervention. Event-based surveillance will also be important, but the focus may shift to current information surrounding a specific outbreak, and the environmental influences that may help or inhibit progress in control of the outbreak. The process of drawing conclusions and making recommendations from various types of surveillance data collected in conjunction with complementary public health information is known as medical intelligence. Medical intelligence is all activities related to early identification of potential health threats, including the process of verifying, assessing, and investigating of disease events to provide actionable public health recommendations for control of an outbreak (Medical intelligence in Europe - European Commission, n.d.).

BIODEFENSE AND OUTBREAK RESPONSE

Societal changes and globalization have led to changes in the risk of both natural disease emergence and reemergence and intentional disease threats. Bioterrorism is the deliberate release of infectious pathogens used to cause illness or death in people, animals, or plants. In infectious disease surveillance, it is important to determine the risk associated with bioterrorism as compared with the risk of natural disease emergence. They may differ in the type of

TABLE 13.2 Event-Based Surveillance Program Systems and Tools

Type of Surveillance Activity	Program	Program Functions	Information Source and Process
Open-source Internet-based reporting system	• Program for Monitoring of Emerging Diseases: ProMED mail (www.promedmail.org)—Federation of American Scientists and SATELLIFE and official program of International Society for Infectious Diseases—was founded in 1994 (ProMED-mail, n.d.)	• Promotes communication among the international infectious disease community • Participants can discuss infectious disease concerns, respond to requests for information, and collaborate on outbreak investigations and prevention efforts.	• Media reports, official reports, online summaries, local observations • Expert human, plant, and animal disease moderators screen, review, and investigate reports • Dissemination to infectious disease network of 70,000 subscribers in at least 185 countries
Real-time surveillance and outbreak monitoring system	• HealthMap(www.healthmap.org) at Boston Children's Hospital—program founded in 2006 (About | HealthMap, n.d.)	• Delivers real-time information on a broad range of emerging infectious diseases to government and private sectors and the public • Achieve a unified and comprehensive view of the current global state of infectious diseases and their effect on human and animal health	• Online news aggregators, eyewitness reports, expert-curated discussions, and validated official reports • Automated system monitoring, organization integration, filtering, visualization, and dissemination in an online maps platform
Internet-based multilingual early warning tool	• Global Public Health Intelligence Network (GPHIN)—Health Canada and World	• Ensure a comprehensive picture of the epidemic threat to global health security	• Systematic event detection • Information and reports from

(Continued)

TABLE 13.2 Event-Based Surveillance Program Systems and Tools—cont'd

Type of Surveillance Activity	Program	Program Functions	Information Source and Process
	Health Organization (WHO)—program founded in 1997	• Identify information about disease outbreaks and other events of potential international public health concern	global media sources such as news wires and websites, outbreak reports from online media and discussion groups • Vetted information from health ministries and WHO offices collaborating centers, and governmental partners.
Internet monitoring and analysis system and information scanning tool	• Medical Information System (MedISys)—(http://medisys.newsbrief.eu/) European Commission (EC) Joint Research Center (JRC) in collaboration with EC Directorate General for Health and Consumer Protection (DG SANCO)—Program Founded 2006	• Provide a science and knowledge tool to aid in identification of current events related to infectious diseases of concern • Rapidly identify potential threats to the public health using information from the Internet • Reinforce the network for surveillance of communicable diseases and the early detection of bioterrorism activities	• Europe Media Monitor (EMM) system uses advanced information extraction techniques to collect information from news sources in over 70 languages • EMM information analyzed for public health relevance and warns users with automatically generated alerts

pathogens used and the mechanism of disease transmission, as well as in presentation and progression. For example, a bioterror attack may present as a point source outbreak; however, the pathogen may be alien to the location or have no plausible mode of natural transmission in that locale (Pavlin, 1999). Using the biosurveillance analytical checklist, it is possible to use epidemic intelligence and epidemiological methods to determine the source of the outbreak and the appropriate mitigation strategies to control the outbreak.

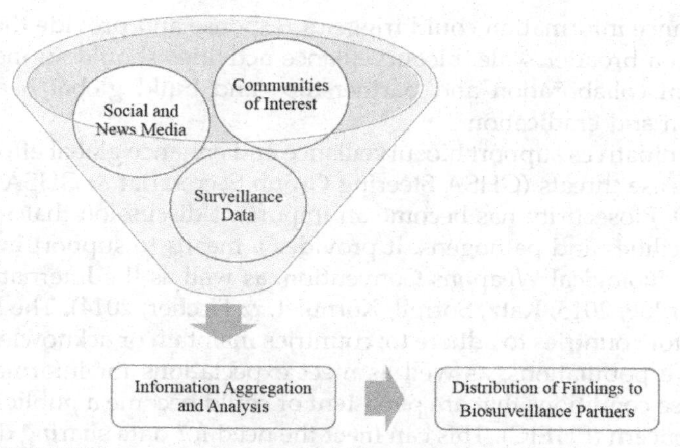

FIGURE 13.1 Information collection and dissemination process.

During outbreak response and control, an important step is determining the at-risk populations and identifying a means for preventing further spread. Depending on the pathogen and the type of disaster or disease event, different processes can be initiated to further enhance surveillance efforts. In some instances, disease registries may be used as a means for monitoring specific sequelae or tracking individuals with an unspecified risk that has some epidemiologic link to the event. For example, after the terror attacks of September 11, 2001, the 9/11 registry was set up to monitor the long-term health impact of environmental exposures for first responders (Li et al., 2012; Nasseri, 2013). Registries can also be useful during an emerging disease outbreak when the potential complications of the disease are unclear or unknown. This was done for the Zika virus outbreak in the Americas in 2016 to monitor the pregnancy complications and birth defects that may occur in Zika-infected mothers and infants (Oduyebo et al., 2016; Russell et al., 2016).

An emerging outbreak support tool is predictive modeling of epidemic progression. Predictive modeling is complex and can provide insight on the scope and projected impact an epidemic may have if no countermeasures are put in place to thwart transmission (Fisman, Khoo, & Tuite, 2014; Nishiura & Chowell, 2014; Nsoesie et al., 2013; Prieto et al., 2012). Although useful, predictive modeling can be counterproductive and raise alarms if the models are not accurate or specific enough to exemplify reality (Roberts, 2013). Despite necessary cautions for reliance on statistical models, applying predictive modeling to outbreak response efforts can support decision-making for interventions and support management of resources throughout the event.

Significance of Partnerships and International Collaboration

Biosurveillance is capable of informing policy and intervention development when recommendations highlight influential factors related to the spread or maintenance of a disease within a population (Kimberly et al., 2014; The White House, 2012). For example, during disasters, surveillance information could trigger a response and provide the basis for mitigation strategies. On a broader scale, biosurveillance activities should support initiatives that foster international collaboration and partnerships and build global health strategies for disease elimination and eradication.

In turn, global initiatives support biosurveillance and enhance global efforts for biosecurity and infectious disease threats (GHSA Steering Group Secretariat & GHSA Preparation Task Force Team, 2015). Biosecurity has become an important discussion that relates to biosafety and security of facilities and pathogens. It provides a means to support international agreements such as the Biological Weapons Convention, as well as the International Health Regulations (IHR) (Burkle, 2015; Katz, Sorrell, Kornblet, & Fischer, 2014). The IHR act as a basic health regulation for countries to adhere to; countries maintain or acknowledge human rights for their respective populations, as well as meet expectations for information sharing and reporting of disease conditions that are persistent or could become a public health emergency of international concern (PHEIC). This can meet the need for data sharing during a PHEIC so that leading SMEs in the field are able to use all available information and data to conduct research and inform policymakers on appropriate actions that need to take place at a country level to develop countermeasures and successfully control the spread of disease.

SUMMARY: PROSPECTIVE BIOSURVEILLANCE

In summary, biosurveillance methods can be used to effectively enhance preparedness in an area prior to a disaster, ensure that postdisaster the systems are in place to identify outbreaks rapidly, provide baseline data to discern urgency of health situations, and forecast potential disease spread or hot spots for disease. Every day there are additional resources developed that can be incorporated into the biosurveillance analytical checklist to provide the most comprehensive pool of information to inform preparedness and response efforts. Advanced protocols in genetic sequencing and molecular epidemiology provide additional insight into expected disease trends and can inform predictive analyses and modeling strategies. Other technological advances such as software applications have been developed to combine the need for increased communication and collaboration with analytical capabilities, geospatial data, and visualization of complex information. Additionally, the growing availability of "big data" from social media, smart phone science, and wearable sensory technology, as well as new initiatives in precision medicine, will shape the evolution of disease detection and epidemiological applications. As these advances continue, it is imperative to maintain the integrity of epidemiological methods while broadening the scope of usefulness of biosurveillance for disaster preparedness and disease prevention.

References

Burkle, F. M. (2015). Global health security demands a strong international health regulations treaty and leadership from a highly resourced World Health Organization. *Disaster Medicine and Public Health Preparedness, 9*(5), 568–580. http://dx.doi.org/10.1017/dmp.2015.26.

Edward, V., Tumacha, A., Kerstin, D., Goran, K., & Tim, E. (2014). Social media and internet-based data in global systems for public health surveillance: A systematic review. *Milbank Quarterly, 92*(1), 7–33.

Fisman, D., Khoo, E., & Tuite, A. (2014). Early epidemic dynamics of the West African 2014 ebola outbreak: Estimates derived with a simple two-parameter model. *PLoS Currents*. http://dx.doi.org/10.1371/currents.outbreaks.89c0d3783f36958d96ebbae97348d571.

GHSA Steering Group Secretariat, GHSA Preparation Task Force Team. (2015). Round-up of GHSA Steering Group and action packages in 2015. *Osong Public Health and Research Perspectives, 6*(6), S28–S33. http://dx.doi.org/10.1016/j.phrp.2015.12.007.

Indicator-based surveillance. (n.d.). Retrieved from http://ecdc.europa.eu/en/activities/surveillance/Pages/index.aspx.

Katz, R., Sorrell, E. M., Kornblet, S. A., & Fischer, J. E. (2014). Global health security agenda and the international health regulations: Moving forward. *Biosecurity and Bioterrorism: Biodefense Strategy, Practice, and Science, 12*(5), 231–238. http://dx.doi.org/10.1089/bsp.2014.0038.

Kimberly, G., Amy, P., Rohit, C., Julie, P., Kevin, R., & Jean-Paul, C. (2014). A review of evaluations of electronic event-based biosurveillance systems. *PLoS One, 9*(10), 1–4. http://dx.doi.org/10.1371/journal.pone.0111222.

Li, J., Cone, J. E., Kahn, A. R., Brackbill, R. M., Farfel, M. R., Greene, C. M., et al. (2012). Association between World Trade Center exposure and excess cancer risk. *JAMA, 308*(23), 2479. http://dx.doi.org/10.1001/jama.2012.110980.

Medical Information System - MEDISYS - Joint Research Centre - European Commission. (n.d.). Retrieved October 2, 2016, from https://ec.europa.eu/jrc/en/scientific-tool/medical-information-system.

Medical intelligence in Europe - European Commission. (n.d.). Retrieved October 2, 2016, from http://ec.europa.eu/health/preparedness_response/generic_preparedness/planning/medical_intelligence_en.htm.

Nasseri, K. (2013). Exposure on September 11, 2001, and cancer risk. *JAMA, 309*(13), 1344. http://dx.doi.org/10.1001/jama.2013.2240.

Nishiura, H., & Chowell, G. (2014). Early transmission dynamics of ebola virus disease (EVD), West Africa, March to August 2014. *Euro Surveillance: Bulletin Européen Sur Les Maladies Transmissibles = European Communicable Disease Bulletin, 19*(36).

Nsoesie, E. O., Brownstein, J. S., Ramakrishnan, N., & Marathe, M. V. (2013). A systematic review of studies on forecasting the dynamics of influenza outbreaks. *Influenza and Other Respiratory Viruses*. http://dx.doi.org/10.1111/irv.12226.

Oduyebo, T., Igbinosa, I., Petersen, E. E., Polen, K. N. D., Pillai, S. K., Ailes, E. C., et al. (2016). Update: Interim guidance for health care providers caring for pregnant women with possible zika virus exposure — United States, July 2016. *Morbidity and Mortality Weekly Report, 65*(29), 739–744. http://dx.doi.org/10.15585/mmwr.mm6529e1.

Pavlin, J. A. (1999). Epidemiology of Bioterrorism. *Emerging Infectious Diseases, 5*(4), 528–530. http://dx.doi.org/10.3201/eid0504.990412.

Prieto, D. M., Das, T. K., Savachkin, A. A., Uribe, A., Izurieta, R., & Malavade, S. (2012). A systematic review to identify areas of enhancements of pandemic simulation models for operational use at provincial and local levels. *BMC Public Health, 12*, 251. http://dx.doi.org/10.1186/1471-2458-12-251.

ProMED-mail. (n.d.). Retrieved October 1, 2016, from http://promedmail.org/aboutus/.

Roberts, M. G. (2013). Epidemic models with uncertainty in the reproduction number. *Journal of Mathematical Biology, 66*(7), 1463–1474. http://dx.doi.org/10.1007/s00285-012-0540-y.

Russell, K., Oliver, S. E., Lewis, L., Barfield, W. D., Cragan, J., Meaney-Delman, D., et al. (2016). Update: Interim guidance for the evaluation and management of infants with possible congenital zika virus infection — United States, August 2016. *Morbidity and Mortality Weekly Report, 65*(33), 870–878. http://dx.doi.org/10.15585/mmwr.mm6533e2.

The White House. (July 31, 2012). *National biosurveillance strategy*. Retrieved from https://obamawhitehouse.archives.gov/sites/default/files/National_Strategy_for_Biosurveillance_July_2012.pdf.

Vignette: Climate Change Effects on Flooding During Hurricane Sandy (2012)

Joel C. Dietrich
NC State University, Raleigh, NC, United States

Hurricane Sandy devastated the Northeast US coastline in 2012. In New York City, it caused power outages that affected nearly 2 million people, forced evacuations of 6500 patients from hospitals and nursing homes, prevented 1.1 million children from attending school for a week, and disrupted the daily travel of about 11 million commuters (Special Initiative for Rebuilding and Resiliency, 2013). Many of these impacts were related to flooding of critical infrastructure (Fig. 1), including nearly 90,000 buildings, and more than $5 billion in damages in the mass transit system (Bernstein, 2013). The maximum observed water level at the tidal gauge located at the southern tip of Manhattan was 5.3 m above the station datum and 2.8 m above the expected tide. This additional water, known as storm surge, was pushed from the open sea by strong winds during the storm. Sandy was one of several recent storms to cause flooding along the US Gulf and Atlantic coasts, including Katrina and Rita (2005), Gustav and Ike (2008), Irene (2011), Isaac (2012), and Hermine and Matthew (2016). Climatic changes are

FIGURE 1 Flooding in Hoboken, New Jersey following Hurricane Sandy. *Credit: Liz Roll/FEMA, 2012.*

Disaster Epidemiology
http://dx.doi.org/10.1016/B978-0-12-809318-4.00020-4

causing these storms to be larger and more intense, last longer, and move farther northward. Their impacts will be more severe to communities in coastal regions in the future.

Climate change has its greatest effect on hurricanes through global warming. Hurricanes act as heat engines (Willoughby, 1999). A hurricane, also known generally as a tropical cyclone, exchanges heat between two reservoirs: the ocean, which is typically warmer in the tropical regions where the cyclone is formed, and the atmosphere, which is much colder at the altitudes near the top of the cyclone. Air is pulled toward the center of the cyclone as it moves over the ocean, picking up water vapor and heat through turbulence at the air–sea interface. This warm air rises near the eyewall, cools as it is pushed outward at high altitudes, and then falls downward to repeat the cycle. Thus hurricanes are sensitive to the sea surface temperature, and they can strengthen quickly when they move over a patch of warmer water. There are several ocean currents that contribute warm water to this process, most notably the Loop Current and Gulf Stream, which connect the warm water from the Caribbean Sea through the Gulf of Mexico and up the eastern coast of the United States. And these waters are becoming warmer. More than 90% of the overall increase in energy due to climate change is stored in the oceans (IPCC, 2013). In regions where tropical cyclones are likely to form, the sea surface temperatures have increased by several tenths of a degree Celsius during the past several decades, due primarily to human-caused changes in greenhouse gases (Santer et al., 2006).

Water expands as it warms, and thus global warming is also a contributor to sea level rise. The global mean sea level has increased by about 1.8 mm/year since 1900, with a larger recent increase by about 3 mm/year since 1993 (Church & White, 2011). This sea level rise is highly variable in location, depending on the local land movement, and it can be larger in regions where the land is shifting downward. In southeastern Virginia, where groundwater withdrawals have caused compaction of aquifers, land subsidence has caused more than half of the relative sea level rise of about 4 mm/year since 1950 (Eggleston & Pope, 2013). These global and local factors contribute to the difficulty in predicting sea level rise through this century. A simple continuation of recent trends would cause an increase in global mean sea level of 17 cm by 2100, and the inclusion of additional warming would further increase this prediction to a range from 28 to 98 cm, depending on the emission scenario (Church et al., 2013). These rising water levels will exacerbate the effects of storm-induced flooding because they provide a higher base level for the storm surge.

This increase in base water level is made worse by an increase in extreme storm frequency. Tropical cyclones are not occurring more often, at least according to the historical record and weather satellite data since the 1970s, but they are becoming more powerful (Frank & Young, 2007). Theory and observations suggest an upper limit on the intensity of tropical cyclones (Bister & Emanuel, 1998), and climate models show an increase in intensity toward this upper limit due to anthropogenic global warming (Knutson & Tuleya, 2004). To analyze these trends, researchers have suggested metrics to combine storm characteristics such as frequency, intensity, and duration. Power dissipation, which is a measure of the potential destructiveness of hurricanes, has more than doubled during the past 30 years (Emanuel, 2005). And storms are moving poleward by 50–60 km per decade (Kossin, Emanuel, & Vecchi, 2014). Thus, when a storm makes landfall in the coastal United States, it is more powerful and more likely to cause flooding than in the past.

These trends converged during Hurricane Sandy. In New York City, the sea level rise during the anthropogenic era has caused the mean flood heights to increase by 1.24 m (Reed et al., 2015), including by 56 cm during the last 250 years (Kemp & Horton, 2013). This increase in base flood level has combined with increases in storm potential to push water toward shore. Flood heights that used to be rare, such as floods that occurred with a probability of once per 100 years, could increase in frequency to occur once every 4 years in the Northeast United States (Frumhoff, McCarthy, Melillo, Moser, & Wuebbles, 2007) and once every 3–20 years in New York City (Lin, Emanuel, Oppenheimer, & Vanmarcke, 2012). During Sandy, the storm surge and flooding were severe due to the extreme size of the storm, which became the largest on record (Blake, Kimberlain, Berg, Cangialosi, & Beven, 2013). Before its landfall on the New Jersey coast, Sandy had tropical-storm- and hurricane-strength winds that extended 778 km and 333 km from its center, respectively (Demuth, Demaria, & Knaff, 2006). These tropical-storm-strength winds had a diameter that was equivalent to the distance from North Carolina to Maine, and they contained more than twice the energy as Hurricane Ike (2008). These winds pushed storm surge against the Northeast US coastline, causing flooding of coastal communities and devastation of critical infrastructure. Hurricane Sandy was an example of recent trends in natural disasters due to climate change, and it emphasizes the need for research and investment in resilient communities.

References

Bernstein, A. (2013). *U.S. Transportation Secretary LaHood announces $3.7 billion in additional Hurricane Sandy disaster relief aid for transit agencies*. U.S. Department of Transportation Document DOT 46–13.

Bister, M., & Emanuel, K. A. (1998). Dissipative heating and hurricane intensity. *Meteorology and Atmospheric Physics, 50*, 233–240.

Blake, E. S., Kimberlain, T. B., Berg, R. J., Cangialosi, J. P., & Beven, J. L., II (2013). *Tropical cyclone report: Hurricane Sandy, (AL182012), 22–29 October 2012*. National Hurricane Center Rep., 157 pp.

Church, J. A., Clark, P. U., Cazenave, A., Gregory, J. M., Jevrejeva, S., Levermann, A., et al. (2013). Sea level change. In T. F. Stocker, D. Qin, G.-K. Plattner, M. Tignor, S. K. Allen, J. Boschung, et al. (Eds.), *Climate change 2013: The physical science basis. Contribution of working group I to the fifth assessment report of the intergovernmental panel on climate change*. Cambridge, United Kingdom and New York, NY, USA: Cambridge University Press.

Church, J. A., & White, N. J. (2011). Sea-level rise from the late 19th to the early 21st century. *Surveys in Geophysics, 32*(4), 585–602.

Demuth, J., Demaria, M., & Knaff, J. A. (2006). Improvement of advanced microwave sounder unit tropical cyclone intensity and size estimation algorithms. *Journal of Applied Meteorology, 45*, 1573–1581.

Eggleston, J., & Pope, J. (2013). Land subsidence and relative sea-level rise in the southern Chesapeake Bay region. *U.S. Geological Survey Circular, 1392*, 30.

Emanuel, K. A. (2005). Increasing destructiveness of tropical cyclones over the past 30 years. *Nature, 436*, 686–688.

Frank, W. M., & Young, G. S. (2007). The interannual variability of tropical cyclones. *Monthly Weather Review, 135*, 3587–3598.

Frumhoff, P. C., McCarthy, J. J., Melillo, J. M., Moser, S. C., & Wuebbles, D. J. (2007). *Confronting climate change in the U.S. Northeast: Science, impacts, and solutions. Synthesis report of the Northeast Climate Impacts Assessment (NECIA)*. Cambridge, MA: Union of Concerned Scientists (UCS).

IPCC. (2013). Summary for policymakers. In T. F. Stocker, D. Qin, G.-K. Plattner, M. Tignor, S. K. Allen, J. Boschung, et al. (Eds.), *Climate Change 2013: The physical science basis. Contribution of working group I to the fifth assessment report of the intergovernmental panel on climate change*. Cambridge, United Kingdom and New York, NY, USA: Cambridge University Press.

Kemp, A. C., & Horton, B. P. (2013). Contribution of relative sea-level rise to historical hurricane flooding in New York City. *Journal of Quaternary Science, 28*(6), 537–541.

Knutson, T. R., & Tuleya, R. E. (2004). Impact of CO2-induced warming on simulated hurricane intensity and precipitation: Sensitivity to the choice of climate model and convective parameterization. *Journal of Climate, 17*, 3477–3495.

Kossin, J. P., Emanuel, K. A., & Vecchi, G. A. (2014). The poleward migration of the location of tropical cyclone maximum intensity. *Nature, 509*, 349–352.

Lin, N., Emanuel, K. A., Oppenheimer, M., & Vanmarcke, E. (2012). Physically based assessment of hurricane surge threat under climate change. *Nature Climate Change, 2*, 462–467.

Reed, A. J., Mann, M. E., Emanuel, K. A., Lin, N., Hortone, B. P., Kemp, A. C., et al. (2015). Increased threat of tropical cyclones and coastal flooding to New York City during the anthropogenic era. *Proceedings of the National Academy of Sciences of the United States of America, 112*(41), 12610–12615.

Santer, B. D., Wigley, T. M. L., Gleckler, P. J., Bonfils, C., Wehner, M. F., AchutaRao, K., et al. (2006). Forced and unforced ocean temperature changes in Atlantic and Pacific tropical cyclogenesis regions. *Proceedings of the National Academy of Sciences of the United States of America, 103*, 13905–13910.

Special Initiative for Rebuilding and Resiliency (SIRR). (2013). *A stronger, more resilient New York*. Available at http://www.nyc.gov/html/sirr/html/report/report.shtml.

Willoughby, H. E. (1999). Hurricane heat engines. *Nature, 401*(6754), 649–650.

Vignette: Disasters and Chronic Medical Conditions

Pamela Allweiss
Centers for Disease Control and Prevention, Atlanta, GA, United States

INTRODUCTION

One of the goals of disaster preparedness planning is to increase the resiliency of vulnerable populations, such as those with preexisting chronic medical conditions. Over the last several decades, there has been an increase in the prevalence of chronic disease or noncommunicable diseases (NCDs) such as heart disease, stroke, cancer, chronic respiratory diseases, and diabetes. For example, as of 2012, about half of all adults living in the United States—117 million people—had at least one chronic health condition. One in four adults had two or more chronic health conditions (Ward, Schiller, & Goodman, 2014).

Disasters can adversely impact people with chronic diseases, causing exacerbations of the illness—even if the illness was controlled prior to the event—that may lead to increased morbidity and mortality. During and after a disaster, people with chronic conditions can experience decreased access to medications and health care, healthy food, clean water and sanitation, access to electricity for air pumps and dialysis equipment, and increased risk of psychological distress, as well as possible exposure to communicable diseases in temporary shelters, zoonosis during floods, and other types of infection due to trauma or environmental living conditions (Arrieta, Foreman, Crook, & Icenogle, 2009; Miller & Arquilla, 2008; Mokdad et al., 2005; Ryan et al., 2015). Some people with NCDs may also have intellectual or developmental disabilities that may hinder evacuation (for instance, Charme, Peacock, Griese, & Howard, 2016). Understanding the needs of people with NCDs before, during, and after a disaster may help responding agencies design plans to increase resiliency in this vulnerable population (Demaio, Jamieson, Horn, de Courten, & Tellier, 2013; Reilly, Degutis, & Morse, 2016; Ryan et al., 2015; Tomio & Sato, 2014; World Health Organization (WHO), 2011).

Disaster Epidemiology
http://dx.doi.org/10.1016/B978-0-12-809318-4.00021-6

Examples of the Impact of Disasters on People With Noncommunicable Diseases

- Hurricane Charley 2004: One-third of households had at least one older adult's medical condition worsen because of the hurricane and one-quarter of households reported that at least one older adult was prevented from receiving routine care for a preexisting condition. Public health officials concluded that, if the rapid needs assessment to evaluate the health status and immediate needs of the affected community had been done earlier, (e.g., 3–5 days after the hurricane) instead of 10–14 days after, the results might have guided decisions to deploy appropriate medications and supplies (Centers for Disease Control and Prevention (CDC), 2004).
- Hurricane Katrina 2005: The percentage of adult evacuees with preexisting chronic conditions was estimated to be 46.3%–58.8%. Issues in providing care for people with NCDs included: decreased availability of medications (Ford et al., 2006; Jhung et al., 2007), poor patient preparedness and lack of self-awareness of their medical conditions and their medications, limited ability to access medical information, poor coordination of aid efforts, and poor communication and collaboration among private and private-aided institutions (Arrieta et al., 2009).
- Great East Japan Earthquake 2011: A review of all papers published about the effect of the earthquake on NCDs found that the most common NCDs addressed were vascular conditions and renal dialysis. These NCDs were probably exacerbated by the cold temperatures at the time of the disaster, dislocation of the population, and the older age of the population (Kako, Arbon, & Mitani, 2014). People with diabetes who were affected by Japan's 2011 triple disaster (earthquake, tsunami, nuclear accident) also experienced a deterioration in their glycemic control (Leppold et al., 2016).
- Hurricane Sandy 2012: A review of the public health consequences of Hurricane Sandy acknowledged that the "second wave" of public health concern was the adverse impact of the storm on people with NCDs due to decreased access to medications and disease management (Reilly et al., 2016). Over 95% of primary care offices in the impacted areas were closed (Sood et al., 2016). Of people who required medical care for NCDs, 19% could not obtain it and 23.3% could not refill a prescription or get needed medical supplies. The number of emergency room visits for medications for NCDs and dialysis peaked in the immediate aftermath of Sandy. Populations with lowest access to care prior to the hurricane had the most problems accessing care during and after hurricane (Davidow et al., 2016).

ACTION STEPS

Given these experiences in past disasters, what can be done to address the needs of those with NCDs before, during, and after disasters? Some examples include the following:

- Know the burden of NCDs in the community (Demaio et al., 2013; Reilly et al., 2016; Tomio & Sato, 2014). For instance, use Behavioral Risk Factor Surveillance System data to determine the prevalence of specific NCDs in your community (Ford et al., 2006).

More information can be found on websites from the Centers for Disease Control and Prevention (CDC) for specific conditions such as *cardiovascular disease, diabetes, cancer, asthma,* and others. Share the information with emergency preparedness planners and response agencies.

- Coordinate educational planning and programs, and include training modules on NCDs for emergency responders. Incorporate preparedness procedures for people with NCDs into existing emergency-related policies, standards, and resources (Demaio et al., 2013).
- Provide training on emergency preparedness for health-care providers who provide routine care for people with NCDs (Demaio et al., 2013; Der-Martirosian et al., 2014; Sphere Project, 2011; Tomio & Sato, 2014; WHO, 2011). Topics could include emergency preparedness, self-management for patients with NCDs or disabilities, as well as trainings for family members and caregivers (Demaio et al., 2013; Motoki et al., 2010). People with NCDs can also prepare a personal evacuation plan or a shelter-in-place kit that includes a week's supply of medications, contact information of family members and physicians, and basic survival supplies (e.g., 3-day supply of food and water, blanket, and flashlight). Training must be provided about the types of food to have in the kit based on medical needs (low sodium, low sugar etc.), and kits must include other supplies such as syringes or glucose monitoring equipment if they have diabetes or extra power supplies if they have a mobility device (FEMA, 2015; Federal Emergency Management Agency (FEMA) and the American Red Cross, 2004).
- Involve multiple stakeholders representing the whole community in disaster planning. Disaster planning should include people with NCDs or disabilities and their families, health-care providers, and public health agencies serving them. Planning can address special issues such as disaster communication and notification, evacuation and emergency transportation, sheltering, medication supplies and equipment, and service animals (CDC, 2015; U.S. Department of Health and Human Services (DHHS), 2016).
- Develop technical guidelines on the clinical management of NCDs in emergencies (Demaio et al., 2013; Ryan et al., 2015; Tomio & Sato, 2014).
- Be aware of local and regional resources such as the CDC-funded chronic disease programs and programs managed by local advocacy groups such as the American Heart Association, American Diabetes Association, and local associations on aging.
- Ensure that people with NCDs have access to essential therapies to reduce morbidity and mortality due to acute complications or exacerbation of their NCDs. Access to medications for NCDs continues to be a problem postdisaster. The Assistant Secretary for Preparedness and Response (ASPR) has sponsored the Prescription Medication Preparedness Initiative to look at barriers and possible solutions, which include strategies to improve access, flexible drug-dispensing policies to help patients build reserves, and training for health-care professionals and patients about disaster planning (Caremeli, Eisenman, Blevins, d'Angona, & Glik, 2013).
- Review the Pandemic and All-Hazards Preparedness Act, which sets the requirements for public health preparedness and response for the U.S. Department of Health and Human Services. This act specifically addresses "individuals who may need additional response assistance" including those who "have disabilities, and chronic medical disorders as being at-risk individuals who need special consideration in planning and response" (U.S. DHHS, 2016).

SUMMARY

Disasters can lead to the exacerbation of NCDs, leading to increased morbidity and mortality. Those working in preparedness (e.g., planners and responders) and chronic disease prevention and treatment (e.g., individuals with NCDs, their families, their health-care providers, and advocacy groups) must collaborate on action steps and training programs that address the needs of this vulnerable population to increase resiliency.

Disclaimer

The findings and conclusions in this report are those of the authors and do not necessarily represent the official position of the Centers for Disease Control and Prevention.
No financial disclosures are reported.

References

Arrieta, M., Foreman, R., Crook, E., & Icenogle, M. L. (2009). Providing continuity of care for chronic diseases in the aftermath of Katrina: From field experience to policy recommendations. *Disaster Medicine and Public Health Preparedness, 3*, 174–182.

Caremeli, K. A., Eisenman, D. P., Blevins, J., d'Angona, B., & Glik, D. C. (2013). Planning for chronic disease medications in disaster: Perspectives from patients, physicians, pharmacists, and insurers. *Disaster Medicine and Public Health Preparedness, 7*(3), 257–265.

Centers for Disease Control and Prevention (CDC). (2004). Rapid assessment of the needs and health status of older adults after hurricane Charley–Charlotte, DeSoto, and Hardee Counties, Florida, August 27–31, 2004. *Morb Mortal Wkly Rep, 53*(837) [PMID:15371964].

Centers for Disease Control and Prevention (CDC). (2015). *Planning for an Emergency: Strategies for identifying and engaging at-risk groups. A guidance document for emergency managers* (1st ed). Atlanta (GA): CDC.

Charme, J., Peacock, G., Griese, S., & Howard, B. (January 1 , 2016). Disaster planning and response with and for people with disabilities. In I. Rubin, D. Greydanus, J. Merrick, & D. Patel (Eds.), *Health care for people with intellectual and developmental disabilities across the Lifespan* (pp. 2237–2246).

Davidow, A., Thomas, P., Kim, S., Passannante, M., Tsai, S., & Tan, C. (2016). Access to care in the wake of hurricane Sandy, New Jersey, 2012. *Disaster Medicine and Public Health Preparedness, 10*, 485–491.

Demaio, A., Jamieson, J., Horn, R., de Courten, M., & Tellier, S. (2013). Non-communicable diseases in emergencies: a call to action. *PLoS Currents, 1*(5) [PMCID: PMC3775888].

Der-Martirosian, C., Riopelle, D., Naranjo, D., Yano, E., Rubenstein, L., & Dobalian, A. (2014). Pre-earthquake burden of illness and post-earthquake health and preparedness in veterans. *Prehospital and Disaster Medicine, 29*(3), 223–229.

Federal Emergency Management Agency (FEMA). (2015). *Preparing makes sense for people with disabilities, others with access and functional needs and the whole community.* Available from https://www.fema.gov/media-library-data/1440775166124-c0fadbb53eb55116746e811f258efb10/FEMA-ReadySpNeeds_web_v3.pdf.

Federal Emergency Management Agency (FEMA) and the American Red Cross. (2004). *Preparing for disaster for people with disabilities and other special needs.* Available from https://www.fema.gov/media-library-data/20130726-1445-20490-6732/fema_476.pdf.

Ford, E. S., Mokdad, A. H., Link, M. W., Garvin, W. S., McGuire, L. C., Jiles, R. B., et al. (April 2006). Chronic disease in health emergencies: In the eye of the hurricane. *Preventing Chronic Disease.* http://www.cdc.gov/pcd/issues/2006/apr/05_0235.htm.

Jhung, M., Shehab, N., Rohr-Allegrini, C., Pollock, D., Sanchez, R., Guerra, F., et al. (2007). Chronic disease and disasters medication demands of Hurricane Katrina evacuees. *Journal of Preventive Medicine, 33*(3), 207–210.

Kako, M., Arbon, P., & Mitani, S. (2014). Literature review on disaster health after the 2011 Great East Japan earthquake. *Prehospital and Disaster Medicine, 29*(1), 54–59.

Leppold, C., Tsubokura, M., Ozaki, A., Nomura, S., Shimada, Y., Morita, T., et al. (2016). Sociodemographic patterning of long-term diabetes mellitus control following Japan's 3.11 triple disaster: A retrospective cohort study. *BMJ Open, 6*, e011455. http://dx.doi.org/10.1136/bmjopen-2016-011455.

Miller, A. C., & Arquilla, B. (2008). Chronic diseases and natural hazards: Impact of disasters on diabetic, renal, and cardiac patients. *Prehospital and Disaster Medicine, 23*(2), 185–194.

Mokdad, A. H., Mensah, G. A., Posner, S. F., Reed, E., Simoes, E. J., & Engelgau, M. (2005). When chronic conditions become acute: Prevention and control of chronic diseases and adverse health outcomes during natural disasters. *Preventing Chronic Disease, 2*, 1–4.

Motoki, E., Mori, K., Kaji, H., Nonami, Y., Fukano, C., Kayano, T., et al. (2010). Development of disaster pamphlets based on health needs of patients with chronic illness. *Prehospital and Disaster Medicine, 25*(4), 354–360.

Office of the Assistant Secretary for Preparedness and Response, & U.S. Department of Health and Human Services. (2016). Prescription medication preparedness initiative – II. In *Collaborative meeting proceedings. March 11, 2016.*

Reilly, M., Degutis, L., & Morse, S. (2016). Investigating the public health impact of Hurricane Sandy. *Disaster Medicine and Public Health Preparedness, 10*(03), 301–303.

Ryan, B., Franklin, R., Burkle, F., Aitken, P., Smith, R., Watt, K., et al. (September 2015). Identifying and describing the impact of cyclone, storm and flood related disasters on treatment management, care and exacerbations of non- communicable diseases and the implications for public health. *PLoS Currents, 28*, 7. currents.dis.62e9286d152de04799644dcca47d9288.

Sood, R., Bocour, A., Kumar, S., Guclu, H., Potter, M., & Shah, T. (2016). Impact on primary care access post-disaster: A case study from the Rockaway Peninsula. *Disaster Med Public Health Preparedness, 10*, 492–495.

Sphere Project. (2011). *Sphere handbook: Humanitarian charter and minimum standards in disaster response, 2011* (3rd ed). Available from:. http://www.spherehandbook.org/en/essential-health-services-non-communicable-diseases-standard-1-non-communicable-diseases/.

Tomio, J., & Sato, H. (2014). Emergency and disaster preparedness for chronically ill patients: A review of recommendations. *Open Access Emergency Medicine, 6*, 69–79.

US Department of Health and Human Services (DHHS). At-risk individuals. Available from http://www.phe.gov/Preparedness/planning/abc/Pages/at-risk.aspx.

Ward, B. W., Schiller, J. S., & Goodman, R. A. (2014). Multiple chronic conditions among US adults: A 2012 update. *Preventing Chronic Disease, 11*, 130389. http://dx.doi.org/10.5888/pcd11.130389.

World Health Organization (WHO). (2011). Disaster risk management for health, non-communicable diseases. *Disaster Risk Management for Health Fact Sheets Global Platform.*

CHAPTER

14

Applications: Disaster Communication and Community Engagement

Jennifer C. Beggs

Michigan Department of Health and Human Services, Lansing, MI, United States

The difference between mere management and leadership is communication. ***Winston Churchill***

INTRODUCTION

Risk communication is defined as a "process of exchanging information among interested parties about the nature, magnitude, significance, or control of a risk" (Covello, 1992, pp. 359–373). Disasters, regardless of nature (e.g., natural, biological, chemical, radiological), will require prompt, accurate, and credible messaging to the public. Best practices for risk communication have evolved over years. Sheppard, Janoske, and Liu (2012) noted that historically, risk communication research tended to most frequently involve case studies and lists of best practices. The focus was typically on organizational risks and reputation response rather than how the public and their behavior was impacted (Heath & O'Hair, 2010). Conceptually, risk communication is often associated with an expected event that examines outcomes from behaviors or exposures (Reynolds & Seeger, 2014). Risk communication can be described as a combination of conflict resolution, public participation, and two-way messaging (Aakko, 2004).

Consistent and sensible messaging is essential to protecting the health of a population affected by a disaster. Crisis and emergency risk communication (CERC) describes the activities of an entity facing an immediate, unexpected emergency and response. CERC principles can be during disaster planning to define the need for messages that describe the current state of events, are informative to take action immediately, occur spontaneously, have short-term preparation, are provided by disaster response leadership, and are community focused (Reynolds & Seeger, 2005). Delivering timely and accurate information in an open dialogue with the audience is essential to building the trust of affected populations.

Disaster Epidemiology
http://dx.doi.org/10.1016/B978-0-12-809318-4.00022-8

FUNDAMENTAL CONCEPTS

Key concepts in the CERC development process include comprehension of CERC principles, identifying the audience, identifying the goal of the message, crafting an accurate and concise message, and delivering the message in a timely fashion by a trusted individual. Tailoring messages to the audience allows for greater dialogue about the disaster. CERC messaging must consider the populace's ethnicity, country of origin, economic status, education level, access to information, literacy, and social standing (Littlefield, 2015). Providing communication in such a way that the recipient can easily understand will build trust and credibility of public officials.

Crisis and Emergency Risk Communication Principles

One of the key principles in CERC preparedness is the "STARCC principle." STARCC stands for simple, timely, accurate, relevant, credible, and consistent (Reynolds, n. d.). By following these concepts throughout the entire disaster response, messengers will be more likely to craft informative statements. Information is limited to three key messages that last approximately 10 seconds or 30 words, are repeated during presentation, use visual graphics such as graphs or slides, and avoid the use of negative words (Steib, 2011). Body language is also a critical component to consider when delivering information to the public. For example, during an interview, a spokesperson should hold good eye contact, sit forward in the chair, keep arms uncrossed, not touch their face, not rest their head in their hand, and speak confidently but not with a raised voice (Steib, 2011). Following these rules will promote credibility among audiences.

Sum of Risk

Steib (2011) states that perception equals reality for the individual affected by a disaster. He further explains risk as Eq. (14.1):

$$\text{Risk} = \text{Hazard} + \text{Outrage} \tag{14.1}$$

An increase in either outrage or hazard will increase risk to the public. CERC can help to minimize outrage in the public by calming fears and diffusing rumors.

Public Information Officer or Spokesperson

Credibility is a critical component when working with the public. The PIO or spokesperson who will be delivering important messages to the public must be reputable and trustworthy.

"Public relations derives from the need for skilled communicators to strategically defend and explain the organization's position in the face of crisis-induced criticism, threat, and uncertainty (Reynolds & Seeger, 2005, p. 46)." For example, during the Fukushima nuclear accident, a complex technical disaster that was precipitated by an earthquake and tsunami, CERC efforts required a strategy based in both scientific evidence and public health ethics.

Public health nurses were discovered to have excellent communication skills and, therefore, contributed greatly to recovery efforts (Shimura, Yamaguchi, Terada, Svendsen, & Kunugita, 2015, p.425). Lastly, the PIO or spokesperson should be ready to answer questions from reporters and journalists. It is critical to work with reporters and journalists so that a story can be pieced together that relays clear and accurate messages. By doing this, the challenge of releasing important information that is not manipulated can be avoided. As Aggergaard (2015) states: "For journalists, the story typically is based on conflict. It features at least one hero and at least one villain and can be pieced together quickly from readily available and usually reliable sources" (p. 13).

RISK COMMUNICATION PLANS

Plans benefit the preparation of message development. The focus of plan writing is to eventually exercise plans to ensure they are comprehensive and functional. Plans should include information such as how the public perceives risk, how the media translate released messages, and address issues each entity may endure (Sinisi, n.d.). CERC predisaster or preparedness planning phases include (1) involving community members, (2) ensuring information comes from multiple channels and is repeated often, and (3) understanding the publics' perceived risk prior to dissemination (Sheppard et al., 2012). Identifying and including these three concepts in a risk communication plan, prior to a disaster, will allow for planning a more developed disaster epidemiology message prior to dissemination. It is also recommended that template messages be developed and attached to the plan (Sutton et al., 2015). Kinder (2012) identifies 10 steps to developing an effective communication plan. This includes the following:

1. Set time aside to develop a plan
2. Train staff on the plan
3. Create a list of worst-case scenarios
4. Create a list of second worst-case scenarios
5. Develop empathetic responses to each scenario
6. Identify public information officers (PIOs) or a spokesperson
7. Develop a dissemination procedure
8. Ensure all staff have a copy of the plan
9. Practice/exercise with focusing on gaps
10. Practice again

Furthermore, plans must be exercised and debriefed as often as possible (Kinder, 2012). Unfortunately, adaption and incorporation of lessons learned from an emergency into the planning process is often neglected (le Roux, 2013).

COLLABORATIONS

To make communications most useful, it is best to include all stakeholders in the message development process. This also ensures that different entities will be less likely to provide divergent messages, reducing the potential for confusion among the populace. Disaster epidemiology communications should include the media, 211 or other call centers, public officials, emergency services representatives, and hospitals. Other stakeholders and materials to consider include television and radio stations, website developers, establishment of a 24-h hotline, fliers in various languages, and predrafting press materials (Steib, 2011).

Ensuring that predisaster relationships exist with the media will encourage collaboration during an actual disaster. One of the most common methods used to disseminate messages is press releases or press conferences. Additional methods to distribute information include telephone meetings, webcasts, emails, broadcast faxes, website postings, and through social media (e.g., Facebook, Twitter) (Reynolds & Seeger, 2014, p. 186). Although it has been estimated that one in five adults have Twitter accounts (Aggergaard, 2015, p. 11), one must remember that Twitter posts must be short and cannot be edited. Credibility can quickly be lost if inaccurate information is released.

It is critical that officials are knowledgeable about the appropriate methods used to provide information to the general public. Therefore, collaborations between local, state, and federal partners are crucial to developing consistent and valuable messages as well. These agencies have subject matter experts that can assist with CERC content.

BARRIERS AND CHALLENGES TO RISK COMMUNICATIONS

Balancing CERC messaging against facts emerging from different entities during a disaster can be a challenge. Barriers to the CERC process include (1) mixed messages from multiple entities, (2) late release of information, (3) demonstrating a paternalistic attitude, (4) not addressing rumors, and (5) power struggles among public officials (Reynolds & Seeger, 2014). In addition, questions from the audience may be difficult to answer. It is recommended that PIOs be familiar with the subject and that they practice answering questions. One way to overcome this challenge is to draft message maps. Answers should be short and focused, and every possible question should be preidentified to practice providing answers. The audience's concerns should be listened to and addressed honestly (Steib, 2011).

Disaster epidemiology typically generates complex data and information. When communicating this information to the public, it is important that technical terms be avoided. Difficult information can often lead to misunderstandings of the message (Chavez, 2011, p. 37). In addition, visual presentation will assist with messaging comprehension. Society has become more visual and appreciates graphs, photos, maps, and illustrations (Morrow, Lazo, Rhome, & Feyen, 2015, p. 38).

Case Study: 2009 H1N1 Influenza A Pandemic

In April of 2009, a new strain of influenza was detected in California. The Centers for Disease Control and Prevention began to research the new strain and identify its characteristics. Cases were soon detected in Texas and New York. Novel influenza A (H1N1) was reported to the World Health Organization (WHO), and the International Health Regulations prompted countries to increase influenza surveillance efforts and stockpiles of antivirals and personnel protective equipment. Vaccine research began immediately. By June 2009, a global influenza pandemic was declared by the WHO (CDC, 2010).

CERC concepts were quickly enacted throughout the United States. Vaccines would take approximately 6 months to produce, and antivirals were only effective for persons developing symptoms. If transmission from person to person could be stopped, resources could be reserved until a vaccine was available and morbidity and mortality would be decreased.

CERC messaging focused on simple action items citizens could take to stay healthy and mitigate risk, such as:

- Wash your hands frequently
- Cover your cough
- Do not aggregate in confined places
- Get your seasonal flu shot
- Stay home if you are sick

A nationwide CERC campaign was launched to reach out to communities and health-care providers. Messages were emphasized in television and radio commercials, posters that businesses and places of faith could display for patrons (Fig. 14.1), bulletins for health-care facilities, newspaper ads, social media such as Twitter and Facebook, wallet cards that could be handed out at health fairs and schools, and letters to parents from schools discussing school dismissals during an outbreak at school. Health officials were constantly interviewed and 211 hotlines were activated. Some states had commercials and public service messages rolling on televisions in the Department of Motor Vehicles while people waited in line. CERC principles were invaluable to controlling the pandemic from spreading further.

FIGURE 14.1 Poster for the 2009 H1N1 influenza A national campaign. *Credit: CDC, 2010.*

Eventually a vaccine was developed and the pandemic was found to be less severe than originally thought. However, using CERC concepts to combat the pandemic may have prevented unnecessary illnesses and could be applied in subsequent years for seasonal flu messaging. The 2009 H1N1 influenza pandemic exemplifies just how important CERC messaging is during a disaster.

CONCLUSIONS

Every disaster can be broken into cycles, including mitigation and prevention, preparedness, response, and recovery (Declan, McFarland, & Clarke, 2014, p. 1). CERC should be utilized during each of the stages to unify public officials and the public by conveying life-saving messages. CERC should not be overlooked as a valuable tool to respond to a disaster. "All type of media play an important and vital role in natural disaster management, early warning systems, mass education of the populace regarding the unpredictable incidents and can help reduce the financial and humanitarian cost of natural disasters" (Ghassabi & Zare-Farashbandi, 2015, p. 5).

Ultimately, CERC applies to disaster epidemiology and public health because they have the responsibility to issue CERC messages. Gaps in which public health is neglected from the CERC practice must be closed. Disaster epidemiology messaging is also an opportune time to bring the community together by providing situational awareness and recommendations to recover from an incident. "Public health agencies need to know whether their messages are relevant for the recipient, reach the most appropriate and targeted groups at risk for injury or death, and are delivered through a modality and device accessible to target populations" (Revere et al., 2015, p. 2). This is the ultimate goal of CERC.

References

Aakko, E. (2004). Risk communication, risk perception, and public health. *Wmj: Official Publication of the State Medical Society of Wisconsin, 103*(1), 25.

Aggergaard, S. (2015). Crisis communications and social media: The game just got together. *The Computer and Internet Lawyer, 32*(8), 11–13.

Centers for Disease Control, Prevention (CDC). (2010). *The 2009 H1N1 Pandemic: Summary Highlights, April 2009–April 2010*. Retrieved from https://www.cdc.gov/h1n1flu/cdcresponse.htm.

Chavez, G. (2011). When numbers aren't enough. *Communication World*. November–December, 37.

Covello, V. T. (1992). Risk communication: An emerging area of health communication research. In S. A. Deetz (Ed.), *Communication yearbook* (Vol. 15, pp. 359–373). Newbury Park, CA: Sage.

Declan, B., McFarland, M., & Clarke, M. (2014). The effectiveness of disaster risk communication: A systematic review of intervention studies. *PLoS Currents, 6*, 1.

Ghassabi, F., & Zare-Farashbandi, F. (2015). The role of media in crisis management: A case study of Azarbayejan earthquake. *International Journal of Health System and Disaster Management*, 3.2, 5.

Heath, R. L., & O'Hair, H. D. (2010). The significance of crisis and risk communications. In R. L. Heath, & H. D. O'Hair (Eds.), *Handbook of risk and crisis communication* (pp. 5–30). New York: Routledge.

Kinder, P. (March 2012). Worst-case scenario: A communications plan can avert disaster. *Utah Business*, 55–58.

le Roux, T. (2013). An exploration of the role of communication during the in-crisis situation. *Journal of Disaster Risk Studies, 5*(2), 5.

Littlefield, R. (2015). Improving how we communicate about infectious disease risks. *Microbe, 10*(7), 281.

Morrow, B., Lazo, J., Rhome, J., & Feyen, J. (January 2015). Improving storm surge risk communication. *American Meteorological Society*, 38.

Revere, D., Calhoun, R., Baseman, J., & Oberle, M. (2015). Exploring bi-directional and SMS messaging for communications between public health agencies and their stakeholders: Aqualitative study. *BioMed Central, 10*, 2.
Reynolds, B. Crisis and emergency risk communication. (n.d.) Retrieved October 10, 2016, from Association of Continuity Professionals website webinar, http://sfba.acp-international.com/files_downloads/CERC_San_Francisco.pdf.
Reynolds, B., & Seeger, M. (2005). Crisis and emergency risk communication as an integrative model. *Journal of Health Communication, 10*, 46.
Reynolds, B., & Seeger, M. (2014). Crisis and emergency risk communication. In *Centers for disease control and prevention publication* (pp. 63–64, 186). Atlanta: GA: Centers for Disease Control and Prevention.
Sheppard, B., Janoske, M., & Liu, B. (2012). *Understanding risk communication theory: A guide for emergency managers and communicators*. Report to Human Factors/Behavioral Sciences Division, Science and Technology Directorate, U.S. Department of Homeland Security. College Park, MD: START, 5, 11.
Shimura, T., Yamaguchi, I., Terada, H., Svendsen, E., & Kunugita, N. (2015). Public health activities for mitigation of radiation exposures and risk communication challenges after the Fukushima nuclear accident. *Journal of Radiation Readiness, 56*(3), 425.
Sinisi, L. Public concerns and risk communications. (n.d.) Retrieved October 10, 2016, from World Health Organization website, p. 185 http://www.who.int/water_sanitation_health/wastewater/wsh0308chap7.pdf.
Steib, P. (2011). Communication in risk situations. *Association of State and Territorial Health Officials*, 6–9.
Sutton, J., Gibson, C., Spiro, E., League, C., Fitzhugh, S., & Butts, C. (2015). What it takes to get passed on: Message contact, style, and structure as predictors of retransmission in the Boston Marathon bombing response. *PLoS One, 10*(8), 16.

Further Reading

Skinner, C., & Rampersad, R. (2014). A revision of communication strategies for effective disaster risk reduction: A case study of the South Durban basin, KwaZulu-Natal, South Africa. *Journal of Disaster Risk Studies, 6*(1), 3.

Vignette From Recent Responses: Roseburg, Oregon Mass Shooting

Akiko M. Saito

Oregon Health Authority – Public Health Division, Portland, OR, United States

On October 1, 2015, on the campus of Umpqua Community College (UCC) in Roseburg, Oregon, a 26-year-old male student walked into his creative writing class, shot and injured nine students, killed eight other students and his professor, and engaged in a shootout with police before taking his own life. Roseburg is a rural town of 22,000, located in Douglas County, with a population of 108,000. At the time of the shooting, UCC had enrollment of 13,000 students including tribal members from the Cow Creek Band of Umpqua Indians. The actual event happened in less than 10 min, but the community response and recovery process that began immediately continued for more than a year after the event. For the state of Oregon, this was the deadliest mass shooting in history. Unfortunately, as we see across the nation, the frequency of these types of events is increasing.

Oregon has a population of 4 million and is a home rule state. This means that each of its 36 counties is responsible for delivering public health and behavioral health to its community. At the state level, the Oregon Health Authority (OHA) provides training, guidance, and funding to the local level for the delivery of public health and behavioral health services. Through the U.S. Health and Human Services (USHHS), Assistant Secretary for Preparedness and Response, and CDC grants, OHA funds local public health authorities and administers a statewide public health emergency preparedness and response program. One of the cornerstones of the statewide public health emergency preparedness program is the administration of the State Emergency Registry of Volunteers in Oregon (SERV-OR). OHA manages SERV-OR, the statewide registry of 2800 health-care volunteers who are licensed, certified, trained, and ready to respond to any public health emergency.

In response to mass casualty events, USHHS has built a national cadre of behavioral health providers who are ready to respond. As the UCC incident unfolded, federal and state behavioral strike teams provided surge capacity for the small rural community whose behavioral health capacity was fragmented and insufficient to meet the increased level of need that the shooting created.

Disaster Epidemiology
http://dx.doi.org/10.1016/B978-0-12-809318-4.00023-X

Immediately following the shooting, a USHHS Behavioral Health team was deployed to UCC with the OHA preparedness and response team in support of the Douglas County community. The USHHS Behavioral Health team consisted of 6 staff members and 11 clinicians, many of whom had responded in the aftermath of the Boston bombing and the Sandy Hook Elementary School shooting. Team members flew from all over the United States and landed in Roseburg within 48 h of the shooting. They spent 2 weeks on campus and in the community, serving students, faculty, and responders with 1826 encounters and collecting data through the use of PsySTART, an evidence-based data collection tool. The team was able to collect PsySTART data and provide real-time metrics to the planning group to make decisions about future surge support. The local planning team determined that there would be a need for additional behavioral health support after the USHHS team completed its 2-week mission and a request went to OHA for an additional SERV-OR deployment.

As the SERV-OR protocol outlines, an alert was sent to 99 behavioral health clinicians through the system and, within 24 h, 12 of the volunteers had returned a positive reply that they were ready and willing to be deployed. One of the main tenets of the SERV-OR system is to ensure that volunteers are fit to respond and, as based on the principles of Emergency Responder Health Monitoring and Surveillance (ERHMS), that after they volunteer, they are fit to return to their home station with any resources needed. Therefore, through a team of local clinical leaders, each of the SERV-OR volunteers' credentials and training were carefully reviewed, and they were individually interviewed by the OHA lead clinician consultant and response staff. A team of nine clinicians was confirmed to serve with one clinician on call. In addition, in partnership with the Veterans Administration, OHA offered a 2-day intensive Cognitive Processing Therapy (CPT) course for 16 local clinicians and 3 SERV-OR volunteers. After the CPT training and a SERV-OR orientation, OHA opened the Umpqua Wellness Center (UWC), a free counseling center for anyone affected by the UCC shooting. The Center was evidence of an effective local and state partnership. A Roseburg private company donated the medical office, the Coordinated Care Organization (http://www.oregon.gov/oha/OHPB/pages/health-reform/ccos.aspx) donated office administrative staff, the state managed and staffed the Center, and community artists donated artwork to create a healing space. The UWC received local media attention and was open for 16 weeks. SERV-OR volunteers staffed the UWC 6 days a week and donated 1722 h to the community.

During the 16-week deployment, OHA monitored the client caseload and held a weekly case conference with all of the SERV-OR clinicians. Each week, data from the Center was shared with OHA leadership, and staff members carefully monitored client usage to make determinations on how long to keep the Center open. As the client loads were reduced and data showed that the need was decreasing, OHA staff members began working on a demobilization plan. After consulting with local clinical leaders and the OHA Privacy Compliance Officer, another SERV-OR volunteer (a former psychiatric nurse practitioner and a current registered nurse with SERV-OR and American Red Cross) was deployed to act as a provider liaison to help in transition of any UWC clients to local providers. This process happened over several weeks and consisted of multiple meetings between the providers and the SERV-OR provider liaison. In the end, 22 clients were successfully transitioned to local providers and their records were transferred securely. Communication went out that the Center was closing, but that the phone number would stay in service so those needing counseling services could still be connected to a local clinician.

OHA staff members instituted Emergency Responder Health Monitoring and Surveillance (ERHMS) principles and held an after-action session for all SERV-OR volunteers and clinical leaders. An after-action facilitator led the discussion for the first half of the day and then local clinicians led individual debriefs with each SERV-OR volunteer to ensure their experience had not created any trauma or additional need for resources. The feedback that was received from the SERV-OR volunteers was that they felt taken care of and appreciated the additional individual support. After the deployment, each of the SERV-OR volunteers was invited to the Governor's Office for a reception. Although there were many gaps identified in the after-action report, it was an overall positive experience for the volunteers. OHA developed an improvement plan based on the SERV-OR clinicians and community feedback. This plan drives the new work of the SERV-OR behavioral health team and a year later, SERV-OR exercised in Roseburg on the 1-year anniversary of the incident. The SERV-OR team, along with Red Cross, local clergy, and clinicians provided counseling services during community-wide events, on campus, at the hospital, and at a drop-in center for responders, hospital staff, students, faculty, and any community members affected by the UCC shooting. OHA provided the overhead administration for the Behavioral Health team and coordinated local efforts around the drop-in centers between September 30 and October 2, 2016. Using ERHMS principles, volunteers were again vetted and provided a security briefing and training orientation before and after each shift.

While we hope that incidents, such as the UCC shooting, will not happen again, we still prepare, exercise, and improve our plans so that in case of an event we can take care of the community in need as well as our emergency responders.

CHAPTER

15

What Can Disaster Epidemiology Contribute to Building Resilient Communities?

Jennifer A. Horney
Texas A&M University, College Station, TX, United States

DEFINING COMMUNITY RESILIENCE

As defined by the National Research Council, resilience is "the ability to prepare and plan for, absorb, recover from, and more successfully adapt to adverse events" (NRC, 2012, p. 1). Rather than waiting for an event to occur and incurring the costs afterward, enhanced resilience allows for better anticipation of disasters and better planning, thereby reducing disaster losses. The types of epidemiologic methods and applications that have been discussed here can contribute to the resilience of individuals and communities to disasters by helping them recover to a new, more resilient normal.

For an example of this, consider the data collected postdisaster via assessments such as Community Assessments for Public Health Emergency Responses (CASPERs). CASPER data can be used in the immediate postdisaster period to identify critical food and water security and sanitation issues to help prevent outbreaks of infectious diseases. Later, these same data, or additional data collected as part of a longitudinal assessment or a preparedness-focused CASPER, can be used to support applications for funding, inform updates to preparedness plans, and guide the development of public education materials about future risks and recommended protective actions so that the impacts of future disaster is reduced (Keim, 2008). However, for disaster epidemiology to significantly contribute to resilience, several challenges must be addressed.

Disaster Epidemiology
http://dx.doi.org/10.1016/B978-0-12-809318-4.00024-1

CHALLENGES TO COLLECTING AND USING EPIDEMIOLOGIC DATA AFTER A DISASTER

After a disaster, various types of data, including surveillance data and data collected as part of rapid assessments and epidemiologic studies, must be rapidly shared with multiple agencies and organizations. While active shelter surveillance or CASPER data may be available in close to real-time, new methods for data collection that allow for real-time two-way communication, such as smartphone applications, may make data available more quickly (Aanensen, Huntley, Feil, & Spratt, 2009). Mobile phone SIM card data can be used to track the location of evacuees to ensure that resources such as shelters, potable water, and health care are available in the location where those affected are moving to (Bengtsson, Lu, Thorson, Garfield, & Von Schreeb, 2011). Beyond formal epidemiologic data, social media and other types of data may be incorporated but should be used with caution as evidence has shown these data may miss certain impacted groups completely or overestimate the intensity of impacts in other groups (Olson, Konty, Paladini, Viboud, & Simonsen, 2013).

In addition to rapidly sharing data, disaster epidemiology data must continue to be communicated to decision-makers, public health colleagues, and the public after a disaster to build resilience. The communication of findings is a step that is often tempting for the disaster epidemiologist to skip; after all, most of the methods and applications we have discussed happen during the response and recovery phases of a disaster. However, epidemiologic data can be important for emergency managers as they update their preparedness plans or develop funding requests related to mitigation activities. For example, a partnership between Texas Department of State Health Services and Bluebonnet Trails Community Services, a local mental health nonprofit who had participated in the development and implementation of the CASPER following the Bastrop Complex Fire in 2011, resulted in a successful application for funding from the Federal Emergency Management Agency for a counseling program, Texas P.R.I.D.E (People Recovering In Spite of Devastating Events) Crisis Counseling (DSHS, 2016). Using CASPER data and other information, Bluebonnet Trails subsequently received funding from the National Institute of Mental Health to develop web-based counseling tools for Bastrop residents impacted by the fire (Hahn Public, 2011). Other funding was received from the Texas General Land Office to address improvements in evacuation routes and from various sources to address longer-term issues such as erosion and reseeding.

Another gap in our ability to use epidemiologic data after a disaster to build resilience is the lack of available baseline and longitudinal data that can be used to identify and compare trends over time. Much of the primary disaster epidemiology data collected focus on a single event and may not be applicable in a future disaster of a different type, scale, or location. Because of this case study—type approach, findings from one study may be difficult to generalize to other disasters or locations. Available public data that could be used to characterize baseline status may not be available at the correct temporal or spatial scale. For example, US census data is only updated every 10 years during the decennial census, while data for many variables are only available at the county level, which may not be a good match spatially for the disaster-impacted area (Flanagan, Gregory, Hallisey, Heitgerd, & Lewis, 2011).

The collection and dissemination of high-quality epidemiologic data after a disaster requires rapid access to human, financial, and other types of resources, such as information technology, infrastructure, or communications. For public health practitioners, competing obligations, such as the need to maintain regular public health services (Committee on Post Disaster Recovery, 2015), and legal requirements, such as compliance and eligibility guidelines for funding (Jordan, 2005) may make it difficult to manage the resource requirements for disaster epidemiology. For academic researchers, the difficulty of rapidly obtaining external funding to support disaster epidemiology activities can reduce their ability to participate. Several efforts are currently underway to address this challenge, including the development of research infrastructures, particularly in the highly vulnerable US Gulf Coast, to support disaster response research and the development of shared processes for rapidly obtaining Institutional Review Board approval for disaster research that balances the need for data with human subjects' protection.

THE IMPORTANCE OF COMMUNITY ENGAGEMENT

Engaging with community stakeholder groups and individual residents living in disaster-impacted neighborhoods around the development of research objectives, survey questions, and other priorities can improve the quality and usefulness of disaster epidemiology data. Including local knowledge in disaster epidemiology projects may make them more relevant to residents, while community partners can serve as a conduit for the distribution of disaster epidemiology findings back to communities and individuals and provide a higher level of assurance that findings will be translated and implemented. However, this type of engagement can be especially challenging when disaster epidemiology surge capacity is provided by outside agencies, such as the Centers for Disease Control and Prevention's Health Studies Branch or Epidemic Intelligence Service, which can be called upon to provide consultation and technical assistance during a disaster response.

Having established relationships with local partners such as policymakers, elected officials, the business community, civic leaders, and the media can assist with gaining and maintaining the confidence and trust of those who have been impacted by the disaster (Jennings, 2008). This can improve response rates, assist in the retention of human subjects for longitudinal assessments, and help mitigate some of the potential structural and cultural barriers to disaster research. Partnerships with a variety of agencies and individuals with different missions and skills can also assist in providing an effective response to the ever-evolving list of threats and disasters. For example, when conducting epidemiological studies after an intentional disaster, such as the inhalational anthrax attacks of 2001, partnerships with law enforcement agencies are necessary to ensure that the interests of both public health and law enforcement are met during the epidemiologic investigation (Goodman, Munson, Dammers, Lazzarini, & Barkley, 2003).

NEXT STEPS

Disaster epidemiology methods can contribute in many ways to the resilience of communities and individuals to the public health and other impacts of disasters. Rapid and ongoing access to high-quality data is essential for the immediate implementation of response and recovery plans and the ongoing task of reducing vulnerability to future disasters. In the short term, making decisions based on deficient or inconsistent data will likely lead to inefficient or inadequate distribution of financial or human resources related to disaster response and crisis management, insufficient availability of trained responders and resources, and a larger burden of both physical and mental health outcomes. In the longer term, the lack of disaster epidemiology data could have impacts on the placement and maintenance of critical infrastructure, such as community health centers or shelters, and the development and implementation of programs to reduce vulnerability. Collecting epidemiologic data to characterize the physical and social vulnerability of communities and individuals will help to strengthen the evidence base for the public health impacts of future disasters.

References

Aanensen, D. M., Huntley, D. M., Feil, E. J., & Spratt, B. G. (2009). EpiCollect: Linking smartphones to web applications for epidemiology, ecology and community data collection. *PLoS One, 4*(9), e6968.

Bengtsson, L., Lu, X., Thorson, A., Garfield, R., & Von Schreeb, J. (2011). Improved response to disasters and outbreaks by tracking population movements with mobile phone network data: A post-earthquake geospatial study in Haiti. *PLoS Medicine, 8*(8), e1001083.

Committee on Post-Disaster Recovery of a Community's Public Health, Medical, and Social Services; Board on Health Sciences Policy; Institute of Medicine. (2015). *Healthy, resilient, and sustainable communities after disasters: Strategies, opportunities, and planning for recovery.* Washington, DC: National Academies Press.

Flanagan, B. E., Gregory, E. W., Hallisey, E. J., Heitgerd, J. L., & Lewis, B. (2011). A social vulnerability index for disaster management. *Journal of Homeland Security and Emergency Management, 8*(1).

Goodman, R. A., Munson, J. W., Dammers, K., Lazzarini, Z., & Barkley, J. P. (2003). Forensic epidemiology: Law at the intersection of public health and criminal investigations. *The Journal of Law, Medicine & Ethics, 31*(4), 684–700.

Hahn Public. (2011). *Bluebonnet trails community services launches new web-based trauma recovery tool. October 20, 2011.* https://www.hahnpublic.com/blog/bluebonnet-trails-community-services-launches-new-web-based-trauma-recovery-tool/.

Jennings, B. (2008). Disaster planning and public health. In M. Crowley (Ed.), *From birth to death and bench to clinic: The hastings center bioethics briefing book for journalists, Policymakers, and Campaigns* (pp. 41–44). Garrison, NY: The Hastings Center.

Jordan, M. Federal Disaster Recovery Programs: Brief Summaries (August 29, 2005, RL31734). U.S Congressional Research Service.

Keim, M. E. (2008). Building human resilience: The role of public health preparedness and response as an adaptation to climate change. *American Journal of Preventive Medicine, 35*(5), 508–516.

National Research Council. (2012). *Disaster resilience: A national imperative.* Washington, DC: The National Academies Press.

Olson, D. R., Konty, K. J., Paladini, M., Viboud, C., & Simonsen, L. (2013). Reassessing google flu trends data for detection of seasonal and pandemic influenza: A comparative epidemiological study at three geographic scales. *PLoS Computational Biology, 9*(10), e1003256.

Texas Department of State Health Services. (2016). *Texas P.R.I.D.E. crisis counseling program: Outreach – support – recovery.* Available at http://www.dshs.texas.gov/mhsa/pride/.

Appendix: Disaster Resources

Suzanne Shurtz

Texas A&M University, College Station, TX, United States

Boston Children's Hospital, HealthMap (2007). **Outbreaks Near Me**. http://www.healthmap.org/outbreaksnearme/

Tracks real-time disease outbreaks on a world map and links to health news. Available as a free app.

Centers for Disease Control and Prevention (CDC) (2014). **Disaster Preparedness and Response: Complete Course. Facilitator Guide**. http://www.cdc.gov/nceh/hsb/disaster/Facilitator_Guide.pdf

A PDF facilitator's guide for an overview of disaster response training. The guide includes learning objectives, estimated completion time, instructional content, practice exercises, and lesson summaries.

Centers for Disease Control and Prevention (CDC) (2016). **Health Studies Branch-Preparedness and Response for Public Health Disasters**. http://www.cdc.gov/nceh/hsb/disaster/

"The Health Studies Branch (HSB) provides expertise and leadership in epidemiology to local, state, federal and international partners to help them prepare for and respond to natural and man-made public health disasters." Includes links to resources, such as a guide book "A Primer for Understanding the Principles and Practices of Disaster Surveillance in the United States."

Centers for Disease Control and Prevention (CDC) (2015). **National Electronic Disease Surveillance System (NEDSS)**. https://wwwn.cdc.gov/nndss/nedss.html

Information exchange for public health surveillance data from healthcare systems to public health departments. Includes the Electronic Laboratory Reporting (ELR) Website.

Centers for Disease Control and Prevention (CDC) (2015). **Resources for Emergency Health Professionals**. http://emergency.cdc.gov/health-professionals.asp

Training and educational materials related to clinical outreach, crisis and risk communication, social media, laboratory information, health alerts, and protecting the health of emergency responders.

Centers for Disease Control and Prevention (CDC), The National Institute for Occupational Safety and Health (NIOSH) (2015). **Emergency Responder Health Monitoring and Surveillance (ERHMS)**. http://www.cdc.gov/niosh/topics/erhms/

Resources for emergency responders to support pre-deployment, deployment and post-deployment, including fact sheets, online training courses, and technical assistance documents.

Council of State and Territorial Epidemiologists (CSTE) (n.d.). **Disaster Epidemiology Tool Repository**. http://cste.site-ym.com/?DisasterEpiToolRep

Disaster response state and agency resources, related to shelter surveillance, syndromic surveillance, mortality surveillance, morbidity surveillance and community health impacts rapid needs assessments.

Council of State and Territorial Epidemiologists (CSTE) (n.d.). **Webinar Library**. http://www.cste.org/?page=WebinarLibrary

View disaster preparedness webinars and webinar slides for public health professionals.

Disaster Resistant Communities Group (2016). **Just In Time Disaster Training Library**. http://www.drc-group.com/project/jitt.html

An online video library related to disaster mitigation, preparedness, response and recovery to support individuals, organizations and agencies.

GIDEON Informatics, Inc. (2016). **Global Infectious Disease and Epidemiology Online Network (GIDEON) Website**. http://www.gideononline.com/

Subscription infectious disease and epidemiology diagnostic and informatics tool. Tracks disease outbreaks. Also available as a subscription app.

National Association of County & City Health Officials (NACCHO) (2011). **Advanced Practice Centers Products**. http://apc.naccho.org/Products/

Browse materials related to emergency preparedness prepared by county and city officials including podcasts, websites, toolkits, presentations. To download some materials may necessitate creating a free account.

Northeast Document Conservation Center (NEDCC) (2006). **dPlan: The Online Disaster-Planning Tool for Cultural and Civic Institutions**. http://www.dplan.org/

An online template for institutions to create their own disaster plan. Includes a demo of an intuitional disaster plan. Register for free to use the template.

Uniformed Services University of the Health Sciences (2016). **National Center for Disaster Medicine & Public Health**. https://ncdmph.usuhs.edu/

A site to provide information and resources related to disaster training, education and educational research in public health preparedness.

U.S. Department of Health & Human Services, National Institutes of Health, National Library of Medicine (2016). **MedlinePlus: Disaster Preparation and Recovery**. https://www.nlm.nih.gov/medlineplus/disasterpreparationandrecovery.html

A consumer-health Website with links to disaster-related guides, videos, games for the general public.

U.S. Department of Health & Human Services, National Institutes of Health, National Library of Medicine (n.d.). **PEOPLE LOCATOR and ReUnite**. https://lpf.nlm.nih.gov/PeopleLocator-ReUnite

A web and downloadable app to assist with finding missing persons after a large disaster. Search using text or photo, report a missing or found person.

U.S. Department of Health & Human Services, National Institutes of Health, National Library of Medicine (2015). **Wireless Information System for Emergency Responders (WISER)**. https://wiser.nlm.nih.gov/

Informatics system for emergency responders dealing with hazardous substances. Includes information on how to identify and contain substances. Also available as a free app.

U.S. Department of Health & Human Services, National Institutes of Health, National Library of Medicine, Disaster Information Management Research Center (2016). **Disaster Information Management Research Center Website**. https://disaster.nlm.nih.gov/

A site compiling and linking to disaster information resources, particularly those provided from the National Library of Medicine.

U.S. Department of Health & Human Services, National Institutes of Health, National Library of Medicine, Disaster Information Management Research Center (2016). **Disaster Lit: Resource Guide for Disaster Medicine and Public Health**. http://disasterlit.nlm.nih.gov/

A resource that searches across disaster-related documents, including reports, conference proceedings, expert guidelines, fact sheets, websites, databases from over 700 organizations.

U.S. Department of Health & Human Services, Office of the Assistant Secretary for Preparedness and Response (2016). **Public Health Emergency**. www.phe.gov/preparedness/

Information on resources and government organizations to support public health emergency response.

Index

Note: 'Page numbers followed by "f" indicate figures, "t" indicate tables and "b" indicate boxes.'

X

Z

Printed in the United States
By Bookmasters

Printed in the United States
By Bookmasters